An Introduction to Composite Materials

An Introduction to Composite Materials

Leo Marshall

MURPHY & MOORE
www.murphy-moorepublishing.com

Published by Murphy & Moore Publishing
1 Rockefeller Plaza,
New York City, NY 10020, USA
www.murphy-moorepublishing.com

An Introduction to Composite Materials
Leo Marshall

International Standard Book Number: 978-1-63987-047-9 (Hardback)

Cataloging-in-Publication Data

An introduction to composite materials / Leo Marshall.
p. cm.
Includes bibliographical references and index.
ISBN 978-1-63987-047-9
1. Composite materials. 2. Materials. I. Marshall, Leo.
TA418.9.C6 I58 2022
620.118--dc23

Contents

Preface

The materials that are produced using two or more constituent materials are known as composite materials. The constituents may or may not have similar physical and chemical properties. Some of the everyday examples of composite materials are plywood, reinforced concrete and fiberglass. Depending upon the structure and materials being used these can be classified into metal matrix composites, ceramic matrix composites, thermoplastic composites, etc. Polyethylene, polyvinylchloride and polyurethane are used extensively as core materials for the formation of composites. The manufacturing of composite materials is done using a wide variety of techniques such as filament welding, lanxide process, z-pinning, tufting and fiber placement processes. Some of the other methods are pressure bag molding, resin transfer molding, braiding, slip forming, continuous casting, etc. This book is a valuable compilation of topics, ranging from the basic to the most complex theories and principles in the field of composite materials. Such selected concepts that redefine composite materials have been presented herein. Those in search of information to further their knowledge will be greatly assisted by this book.

To facilitate a deeper understanding of the contents of this book a short introduction of every chapter is written below:

Chapter 1- A material which is produced using two or more different materials is known as a composite material. Some of the major properties of composites are isostrain rules of mixtures, isostress rule of mixtures and coefficient of thermal expansion. This chapter has been carefully written to provide an easy introduction to these facets of composite materials.

Chapter 2- A few of the different types of matrix composites are metal matrix composite, ceramic matrix composite, polymer matrix composite and carbon matrix composite. A metal matrix composite has at least one constituent part which is a metal. The topics elaborated in this chapter will help in gaining a better perspective about these types of matrix composites.

Chapter 3- There are various techniques which can be used for the fabrication of composite materials such as open moulding, closed moulding, autoclave moulding, lanxide process, Z-pinning, stitching technology and tufting. This chapter closely examines these fabrication techniques to provide an extensive understanding of the subject.

Chapter 4- A multiphase solid material wherein at least one phase has a dimension of less than 100 nanometers is known as a nanocomposite. A few of the major types of nanocomposites are polymer nanocomposite, carbon nanocomposite, nanowire and nanocellulose. This chapter discusses in detail these types of nanocomposites.

Chapter 5- Composite materials find application in a variety of fields. They have automobile applications, aerospace applications, marine applications, biomedical applications, sports applications, civil applications and electrical applications. This chapter closely examines these key applications of composites to provide an extensive understanding of the subject.

I owe the completion of this book to the never-ending support of my family, who supported me throughout the project.

Leo Marshall

1

Composite Materials: An Introduction

A material which is produced using two or more different materials is known as a composite material. Some of the major properties of composites are isostrain rules of mixtures, isostress rule of mixtures and coefficient of thermal expansion. This chapter has been carefully written to provide an easy introduction to these facets of composite materials.

After making and controlling fire and inventing the wheel, spinning of continuous yarns is probably the most important development of mankind, enabling him to survive outside the tropical climate zones and spread across the surface of the Earth. Flexible fabrics made of locally grown and spun fibres as cotton; flax and jute were a big step forward compared to animal skins. More and more natural resources were used, soon resulting in the first composites; straw reinforced walls, and bows and chariots made of glued layers of wood, bone and horn. More durable materials as wood and metal soon replaced these antique composites.

A composite material can be defined as a combination of two or more materials that results in better properties than those of the individual components used alone. In contrast to metallic alloys, each material retains its separate chemical, physical, and mechanical properties. The two constituents are a reinforcement and a matrix. The main advantages of composite materials are their high strength and stiffness, combined with low density, when compared with bulk materials, allowing for a weight reduction in the finished part.

The reinforcing phase provides the strength and stiffness. In most cases, the reinforcement is harder, stronger, and stiffer than the matrix. The reinforcement is usually a fiber or a particulate. Particulate composites have dimensions that are approximately equal in all directions. They may be spherical, platelets, or any other regular or irregular geometry. Particulate composites tend to be much weaker and less stiff than continuous fiber composites, but they are usually much less expensive. Particulate reinforced composites usually contain less reinforcement (up to 40 to 50 volume percent) due to processing difficulties and brittleness.

A fiber has a length that is much greater than its diameter. The length-to-diameter (l/d) ratio is known as the aspect ratio and can vary greatly. Continuous fibers have long aspect ratios, while discontinuous fibers have short aspect ratios. Continuous-fiber composites normally have a preferred orientation, while discontinuous fibers generally have a random orientation. Examples of continuous reinforcements include unidirectional, woven cloth, and helical winding, while examples of discontinuous reinforcements are chopped fibers and random mat. Continuous-fiber composites are often made into laminates by stacking single sheets of continuous fibers in different orientations to obtain the desired strength and stiffness properties with fiber volumes as high as 60 to 70 percent. Fibers produce high-strength composites because of their small diameter; they contain far fewer defects (normally surface defects) compared to the material produced in bulk. As a general rule, the smaller the diameter of the fiber, the higher its strength, but often the cost increases as the diameter becomes smaller. In addition, smaller-diameter high-strength fibers have greater flexibility and are more amenable to fabrication processes such as weaving or forming over radii. Typical fibers include glass, aramid, and carbon, which may be continuous or discontinuous.

Types of Materials

Materials can be classified as either isotropic or anisotropic. Isotropic materials have the same material properties in all directions, and normal loads create only normal strains. By comparison, anisotropic materials have different material properties in all directions at a point in the body. There are no material planes of symmetry, and normal loads create both normal strains and shear strains. A material is isotropic if the properties are independent of direction within the material.

For example, consider the element of an isotropic material shown in Figure below. If the material is loaded along its 0°, 45°, and 90° directions, the modulus of elasticity (E) is the same in each direction ($E_{0°} = E_{45°} = E_{90°}$). However, if the material is anisotropic, it has properties that vary with direction within the material. In this example, the moduli are different in each direction ($E_{0°} \neq E_{45°} \neq E_{90°}$). While the modulus of elasticity is used in the example, the same dependence on direction can occur for other material properties, such as ultimate strength, Poisson's ratio, and thermal expansion coefficient.

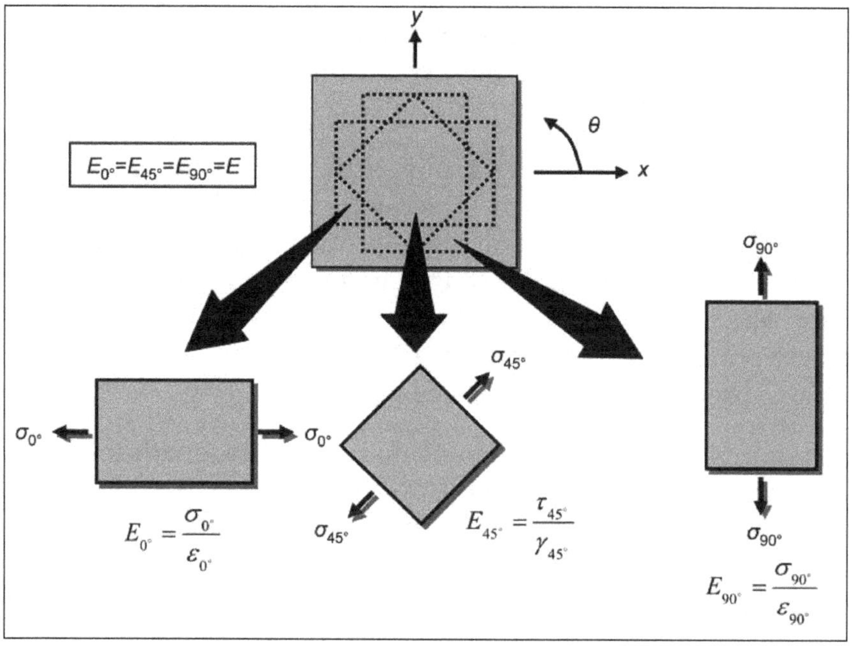

Element of isotropic material under stress.

Bulk materials, such as metals and polymers, are normally treated as isotropic materials, while composites are treated as anisotropic. However, even bulk materials such as metals can become anisotropic–for example, if they are highly cold worked to produce grain alignment in a certain direction.

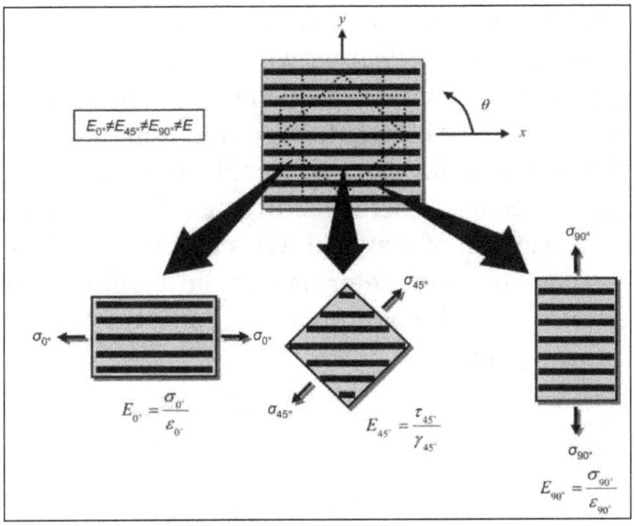

Element of composite ply material under stress.

Consider the unidirectional fiber-reinforced composite ply (also known as a lamina) shown in Figure below. The coordinate system used to describe the ply is labeled the 1-2-3 axes. In this case, the 1-axis is defined to be parallel to the fibers (0°), the 2-axis is defined to lie within the plane of the plate and is perpendicular to the fibers (90°), and the 3-axis is defined to be normal to the plane of the plate. The 1-2-3 coordinate system is referred to as the principal material coordinate system. If the plate is loaded parallel to the fibers (one- or zero-degree direction), the modulus of elasticity E_{11} approaches that of the fibers. If the plate is loaded perpendicular to the fibers in the two- or 90-degree direction, the modulus E_{22} is much lower, approaching that of the relatively less stiff matrix. Since $E_{11} >> E_{22}$ and the modulus varies with direction within the material, the material is anisotropic.

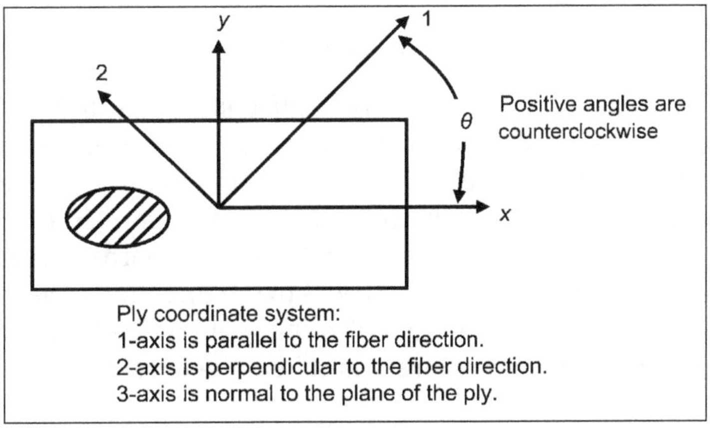

Ply angle definition.

Composites are a subclass of anisotropic materials that are classified as orthotropic. Orthotropic materials have properties that are different in three mutually perpendicular directions. They have three mutually perpendicular axes of symmetry, and a load applied parallel to these axes produces only normal strains. However, loads that are not applied parallel to these axes produce both normal and shear strains. Therefore, orthotropic mechanical properties are a function of orientation.

Consider the unidirectional composite shown in the upper portion of Fig. 1.7, where the unidirectional fibers are oriented at an angle of 45 degrees with respect to the x-axis. In the small, isolated square element from the gage region, because the element is initially square (in this example), the fibers are parallel to diagonal AD of the element. In contrast, fibers are perpendicular to diagonal BC. This implies that the element is stiffer along diagonal AD than along diagonal BC. When a tensile stress is applied, the square element deforms. Because the stiffness is higher along diagonal AD than along diagonal BC, the length of diagonal AD is not increased as much as that of diagonal BC. Therefore, the initially square element deforms into the shape of a parallelogram. Because the element has been distorted into a parallelogram, a shear strain γxy is induced as a result of coupling between the axial strains e_{xx} and e_{yy}.

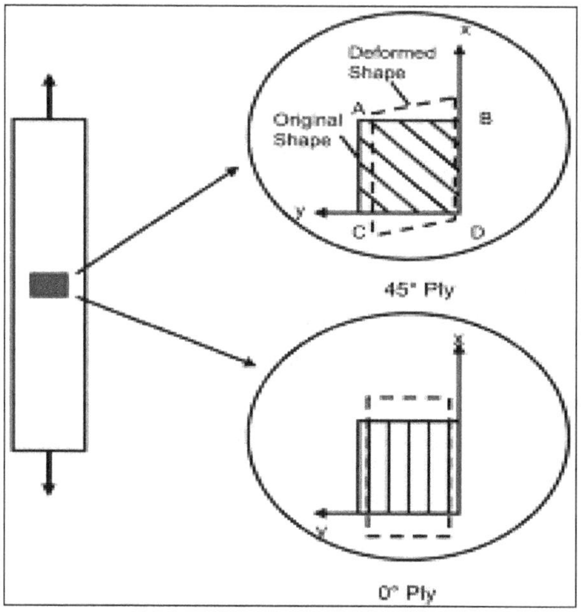

Shear coupling in a 45° ply.

If the fibers are aligned parallel to the direction of applied stress, as in the lower portion of Figure above, the coupling between e_{xx} and e_{yy} does not occur. In this case, the application of a tensile stress produces elongation in the x-direction and contraction in the y-direction, and the distorted element remains rectangular. Therefore, the coupling effects exhibited by composites occur only if stress and strain are referenced to a non– principal material coordinate system. Thus, when the fibers are aligned parallel (0°) or perpendicular (90°) to the direction of applied stress, the lamina is known as a specially orthotropic lamina (θ = 0° or 90°). A lamina that is not aligned parallel or perpendicular to the direction of applied stress is called a general orthotropic lamina (θ ≠ 0° or 90°).

Laminates

When there is a single ply or a lay-up in which all of the layers or plies are stacked in the same orientation, the lay-up is called a lamina. When the plies are stacked at various angles, the lay-up is called a laminate. Continuous-fiber composites are normally laminated materials (Fig. 1.8) in which the individual layers, plies, or laminae are oriented in directions that will enhance the strength in the primary load direction.

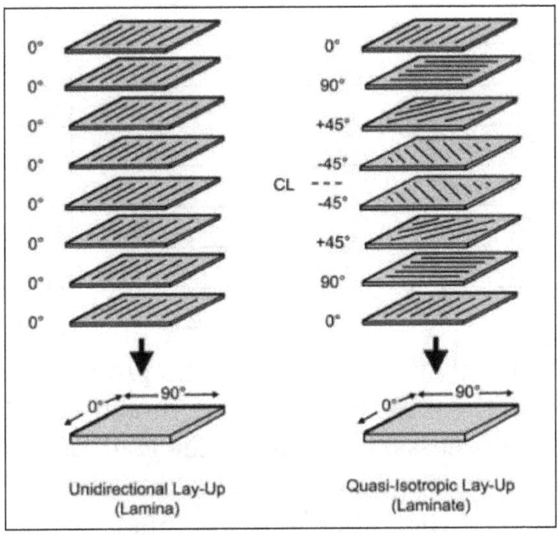

Lamina and laminate lay-ups.

Unidirectional (0°) laminae are extremely strong and stiff in the 0° direction. However, they are very weak in the 90° direction because the load must be carried by the much weaker polymeric matrix. While a high-strength fiber can have a tensile strength of 500 ksi (3500 MPa) or more, a typical polymeric matrix normally has a tensile strength of only 5 to 10 ksi (35 to 70 MPa).

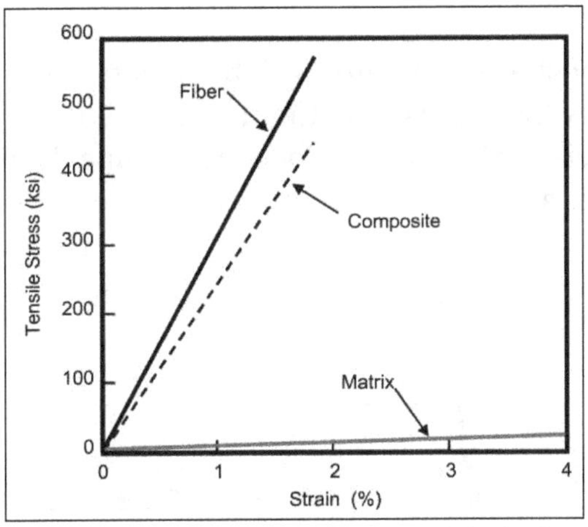

Comparison of tensile properties of fiber, matrix, and composite.

The longitudinal tension and compression loads are carried by the fibers, while the matrix distributes the loads between the fibers in tension and stabilizes the fibers and prevents them from buckling in compression. The matrix is also the primary load carrier for interlaminar shear (i.e., shear between the layers) and transverse (90°) tension. The relative roles of the fiber and the matrix in detemining mechanical properties are summarized in table.

Table: Effect of fiber and matrix on mechanical properties.

	Dominating composite constituent	
Mechanical property	Fiber	Matrix
1. Unidirectional		
0° tension	√	...

0° compression	√	√
Shear	...	√
90° tension		√
2. Laminate		
Tension	√	...
Compression	√	√
In-plane shear	√	√
Interlaminar shear	...	√

Because the fiber orientation directly impacts mechanical properties, it seems logical to orient as many of the layers as possible in the main load-carrying direction. While this approach may work for some structures, it is usually necessary to balance the load-carrying capability in a number of different directions, such as the 0°, +45°, -45°, and 90° directions. Figure shows a photomicrograph of a cross-plied continuous carbon fiber/epoxy laminate. A balanced laminate having equal numbers of plies in the 0°, +45°, −45°, and 90° degrees directions is called a quasi-isotropic laminate, because it carries equal loads in all four directions.

Reinforcements

Reinforcements for the composites can be fibers, fabrics particles or whiskers. Fibers are essentially characterized by one very long axis with other two axes either often circular or near circular. Particles have no preferred orientation and so does their shape. Whiskers have a preferred shape but are small both in diameter and length as compared to fibers. Figure below shows types of reinforcements in composites.

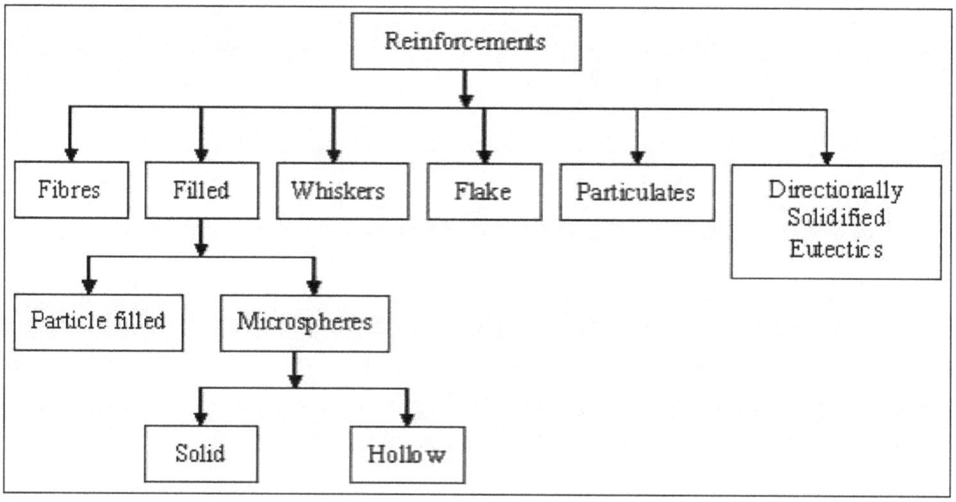

Reinforcements

Reinforcing constituents in composites, as the word indicates, provide the strength that makes the composite what it is. But they also serve certain additional purposes of heat resistance or conduction, resistance to corrosion and provide rigidity. Reinforcement can be made to perform all or one of these functions as per the requirements.

A reinforcement that embellishes the matrix strength must be stronger and stiffer than the matrix and capable of changing failure mechanism to the advantage of the composite. This means that the ductility should be minimal or even nil the composite must behave as brittle as possible.

Types of Reinforcements

Fiber Reinforced Composites

The orientation of the fiber in the matrix is an indication of the strength of the composite and the strength is greatest along the longitudinal directional of fiber. This doesn't mean the longitudinal fibers can take the same quantum of load irrespective of the direction in which it is applied. Optimum performance from longitudinal fibers can be obtained if the load is applied along its direction. The slightest shift in the angle of loading may drastically reduce the strength of the composite.

Unidirectional loading is found in few structures and hence it is prudent to give a mix of orientations for fibers in composites particularly where the load is expected to be the heaviest. Monolayer tapes consisting of continuous or discontinuous fibers can be oriented unidirectional stacked into plies containing layers of filaments also oriented in the same direction. More complicated orientations are possible too and nowadays, computers are used to make projections of such variations to suit specific needs. In short, in planar composites, strength can be changed from unidirectional fiber oriented composites that result in composites with nearly isotropic properties.

Properties of angle-plied composites which are not quasi-isotropic may vary with the number of plies and their orientations. Composite variables in such composites are assumed to have a constant ratio and the matrices are considered relatively weaker than the fibers. The strength of the fiber in any one of the three axes would, therefore be one-third the unidirectional fiber composite, assuming that the volume percentage is equal in all three axes.

However, orientation of short fibers by different methods is also possible like random orientations by sprinkling on to given plane or addition of matrix in liquid or solid state before or after the fiber deposition. Even three-dimensional orientations can achieve in this way.

There are several methods of random fiber orientations, which in a two-dimensional one, yield composites with one-third the strength of a unidirectional fiber-stressed composite, in the direction of fibers. In a 3-dimension, it would result in a composite with a comparable ratio, about less than one-fifth.

In very strong matrices, moduli and strengths have not been observed. Application of the strength of the composites with such matrices and several orientations is also possible. The longitudinal strength can be calculated on the basis of the assumption that fibers have been reduced to their effective strength on approximation value in composites with strong matrices and non-longitudinally orientated fibers.

It goes without saying that fiber composites may be constructed with either continuous or short fibers. Experience has shown that continuous fibers (or filaments) exhibit better orientation, although it does not reflect in their performance. Fibers have a high aspect ratio, i.e., their lengths being several times greater than their effective diameters. This is the reason why filaments are manufactured using continuous process. This finished filaments.

Mass production of filaments is well known and they match with several matrices in different ways like winding, twisting, weaving and knitting, which exhibit the characteristics of a fabric. Since they have low densities and high strengths, the fiber lengths in filaments or other fibers yield considerable influence on the mechanical properties as well as the response of composites to processing and procedures. Shorter fibers with proper orientation composites that use glass, ceramic or multi-purpose fibers can be endowed with considerably higher strength than those that use continuous fibers. Short fibers are also known to their theoretical strength. The continuous fiber constituent of a composite is often joined by the filament winding process in which the matrix impregnated fiber wrapped around a mandrel shaped like the part over which the composite is to be placed, and equitable load distribution and favorable orientation of the fiber is possible in the finished product. However, winding is mostly confined to fabrication of bodies of revolution and the occasional irregular, flat surface.

Short-length fibers incorporated by the open- or close-mould process are found to be less efficient, although the input costs are considerably lower than filament winding. Most fibers in use currently are solids which are easy to produce and handle, having a circular cross-section, although a few non-conventional shaped and hollow fibers show signs of capabilities that can improve the mechanical qualities of the composites.

Laminar Composites

Laminar Composites are found in as many combinations as the number of materials. They can be described as materials comprising of layers of materials bonded together. These may be of several layers of two or more metal materials occurring alternately or in a determined order more than once, and in as many numbers as required for a specific purpose.

Clad and sandwich laminates have many areas as it ought to be, although they are known to follow the rule of mixtures from the modulus and strength point of view. Other intrinsic values pertaining to metal-matrix, metal-reinforced composites are also fairly well known.

Powder metallurgical processes like roll bonding, hot pressing, diffusion bonding, brazing and so on can be employed for the fabrication of different alloys of sheet, foil, powder or sprayed materials. It is not possible to achieve high strength materials unlike the fiber version. But sheets and foils can be made isotropic in two dimensions more easily than fibers. Foils and sheets are also made to exhibit high percentages of which they are put. For instance, a strong sheet may use over 92% in laminar structure, while it is difficult to make fibers of such compositions. Fiber laminates cannot over 75% strong fibers.

The main functional types of metal-metal laminates that do not posses high strength or stiffness are single layered ones that endow the composites with special properties, apart from being costeffective. They are usually made by pre-coating or cladding methods.

Pre-coated metals are formed by forming by forming a layer on a substrate, in the form of a thin continuous film. This is achieved by hot dipping and occasionally by chemical plating and electroplating. Clad metals are found to be suitable for more intensive environments where denser faces are required.

There are many combinations of sheet and foil which function as adhesives at low temperatures. Such materials, plastics or metals, may be clubbed together with a third constituent. Pre-painted or pre-finished metal whose primary advantage is elimination of final finishing by the user is the best known metal-organic laminate. Several combinations of metal-plastic, vinyl-metal laminates, organic films and metals, account for upto 95% of metal-plastic laminates known. They are made by adhesive bonding processes.

Particulate Reinforced Composites

Microstructures of metal and ceramics composites, which show particles of one phase strewn in the other, are known as particle reinforced composites. Square, triangular and round shapes of reinforcement are known, but the dimensions of all their sides are observed to be more or less equal. The size and volume concentration of the dispersoid distinguishes it from dispersion hardened materials.

The dispersed size in particulate composites is of the order of a few microns and volume concentration is greater than 28%. The difference between particulate composite and dispersion strengthened ones is, thus, oblivious. The mechanism used to strengthen each of them is also different. The dispersed in the dispersion-strengthen materials reinforces the matrix alloy by arresting motion of dislocations and needs large forces to fracture the restriction created by dispersion.

In particulate composites, the particles strengthen the system by the hydrostatic coercion of fillers in matrices and by their hardness relative to the matrix. Three-dimensional reinforcement in composites offers isotropic properties, because of the three systematical orthogonal planes. Since it is not homogeneous, the material properties acquire sensitivity to the constituent properties, as well as the interfacial properties and geometric shapes of the array. The composite's strength usually depends on the diameter of the particles, the inter-particle spacing, and the volume fraction of the reinforcement. The matrix properties influence the behaviour of particulate composite too.

Whiskers

Single crystals grown with nearly zero defects are termed whiskers. They are usually discontinuous and short fibers of different cross sections made from several materials like graphite, silicon carbide, copper, iron etc. Typical lengths are in 3 to 55 N.M. ranges. Whiskers differ from particles in that, whiskers have a definite length to width ratio greater than one. Whiskers can have extraordinary strengths upto 7000 MPa.

Whiskers were grown quite incidentally in laboratories for the first time, while nature has some geological structures that can be described as whiskers. Initially, their usefulness was overlooked as they were dismissed as incidental by-products of other structure. However, study on crystal structures and growth in metals sparked off an interest in them, and also the study of defects that affect the strength of materials, they came to be incorporated in composites using several methods, including power metallurgy and slip-casting techniques.

Metal-whisker combination, strengthening the system at high temperatures, has been demonstrated at the laboratory level. But whiskers are fine, small sized materials not easy to handle and this comes in the way of incorporating them into engineering materials to come out with a superior quality composite system.

Early research has shown that whisker strength varies inversely with effective diameter. When whiskers were embedded in matrices, whiskers of diameter upto 2 to 10μm yielded fairly good composites. Ceramic material's whiskers have high moduli, useful strengths and low densities.

Specific strength and specific modulus are very high and this makes ceramic whiskers suitable for low weight structure composites. They also resist temperature, mechanical damage and oxidation more responsively than metallic whiskers, which are denser than ceramic whiskers. However, they are not commercially viable because they are damaged while handling.

Matrix Materials

Matrix is the continuous phase which is a polymer, metal, or ceramic. Polymers have low strength and stiffness, metals have intermediate strength and stiffness but high ductility, and ceramics have high strength and stiffness but are brittle. The matrix (continuous phase) performs several critical functions, including maintaining the fibers in the proper orientation and spacing and protecting them from abrasion and the environment. In polymer and metal matrix composites that form a strong bond between the fiber and the matrix, the matrix transmits loads from the matrix to the fibers through shear loading at the interface. In ceramic matrix composites, the objective is often to increase the toughness rather than the strength and stiffness; therefore, a low interfacial strength bond is desirable.

The type and quantity of the reinforcement determine the final properties. Figure below shows that the highest strength and modulus are obtained with continuous-fiber composites. There is a practical limit of about 70 volume percent reinforcement that can be added to form a composite. At higher percentages, there is too little matrix to support the fibers effectively. The theoretical strength of discontinuous-fiber composites can approach that of continuous-fiber composites if their aspect ratios are great enough and they are aligned, but it is difficult in practice to maintain good alignment with discontinuous fibers. Discontinuous-fiber composites are normally somewhat random in alignment, which dramatically reduces their strength and modulus. However, discontinuous-fiber composites are generally much less costly than continuous-fiber composites. Therefore, continuous-fiber composites are used where higher strength and stiffness are required (but at a higher cost), and discontinuous-fiber composites are used where cost is the main driver and strength and stiffness are less important.

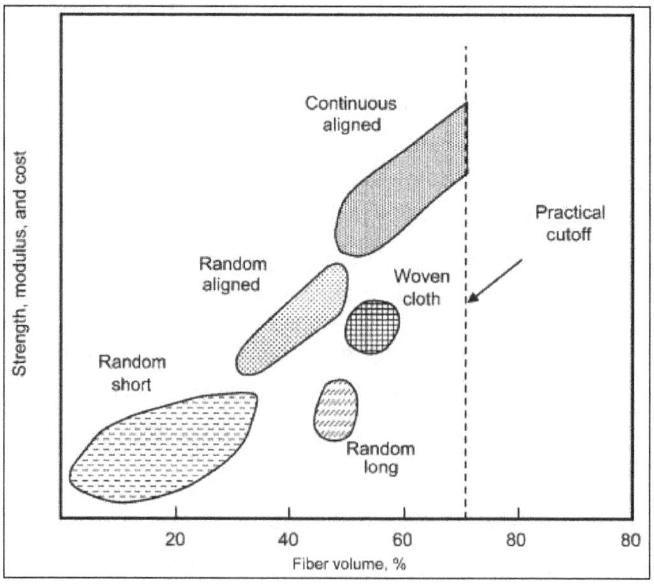

Influence of reinforcement type and quantity on composite performance.

Both the reinforcement type and the matrix affect processing. The major processing routes for polymer matrix composites are shown in Figure below. Two types of polymer matrices are shown: thermosets and thermoplastics. A thermoset starts as a low-viscosity resin that reacts and cures during processing, forming an intractable solid. A thermoplastic is a high-viscosity resin that is processed by heating it above its melting temperature. Because a thermoset resin sets up and cures during processing, it cannot be reprocessed by reheating. By comparison, a thermoplastic can be reheated above its melting temperature for additional processing. There are processes for both classes of resins that are more amenable to discontinuous fibers and others that are more amenable to continuous fibers. In general, because metal and ceramic matrix composites require very high temperatures and sometimes high pressures for processing, they are normally much more expensive than polymer matrix composites. However, they have much better thermal stability.

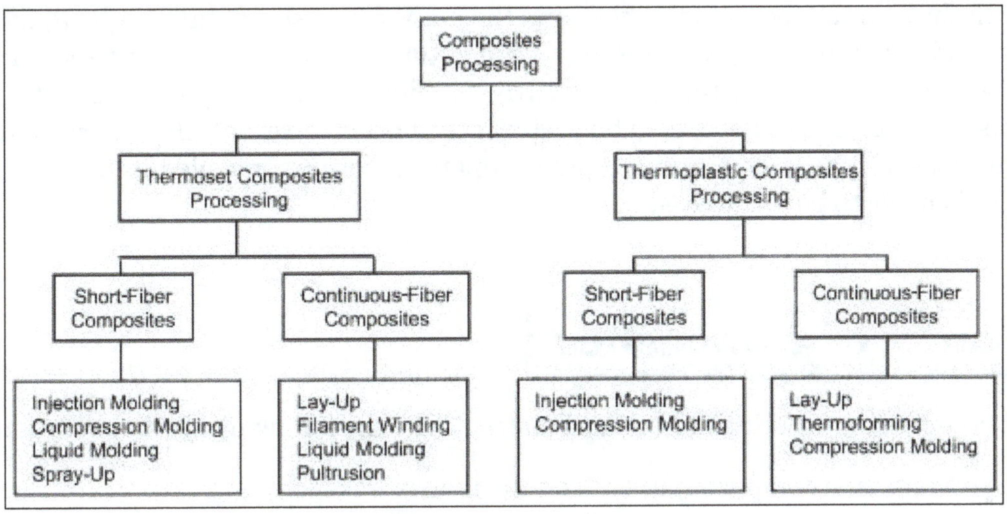

Major polymer matrix composite fabrication processes.

Interphase

The primary dimension of any composite is the interface. The interface is the area where the different materials in a composite coincide. In order to have a successful, applicable composite, one must form an interface that is strong and favorable towards maximum compatibility. A good interface is imperative for a material to survive under stress.

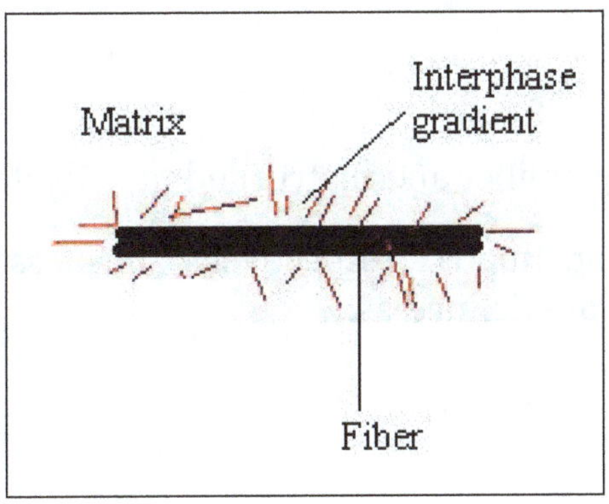

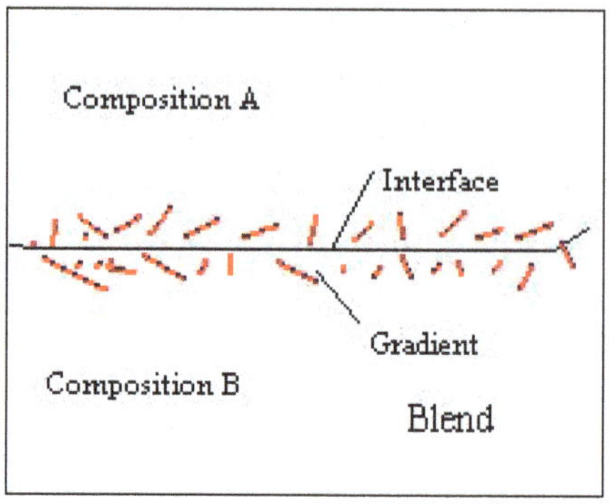

In viewing the above graphic of a fiber reinforced matrix, one can see the presence of a medium that has both the characteristics of the fiber and the matrix. The size of this gradient, the chemical interaction, and the number of gradients present in a composite determine the strength and application of the material. The same concept holds true for blends. The interphase created between the components must achieve some level of favorable interaction to prevent the materials from "keeping to themselves" or phase separating. If phase separation occurs in a blend, then a combination of properties is likely to be unsuccessful.

Mechanics of Composites

Consider a composite consisting of fiber and matrix. Take the following symbol notations:

$V_{c,f,m}$ = volume of composite, fiber, and the matrix, respectively.

$P_{c,f,m}$ = density of composite, fiber and the metrix, respectively.

Now define the fiber volume fraction. V_f and the matrix volume fraction V_m as:

$$V_f = \frac{v_f}{v_c}, \quad V_m = \frac{v_m}{v_c},$$

The sum of volume fractions is:

$$V_f + V_m = 1, \quad v_f + v_m = v_c.$$

Mass Fractions

Consider a composite consisting of fiber and matrix and take the following symbol notation:

$w_{c,f,m}$= mass of composite, fiber, and matrix, respectively. The mass fraction (weight fraction) of the fibers (W_f) and the matrix (W_m) are defined as:

$$W_f = \frac{w_f}{w_c}$$

and,

$$W_m = \frac{w_m}{w_c}$$

Note that the sum of mass fractions is:

$$W_f + W_m = 1, \quad w_f + w_m = w_c.$$

From the definition of the density of a single material:

$$w_c = r_c v_c$$
$$w_f = r_f v_f$$
$$w_m = r_m v_m$$

Substituting above equations, the mass fractions and volume fractions are related as:

$$W_f = \frac{\rho_f}{\rho_c} V_f$$
$$W_m = \frac{\rho_m}{\rho_c} V_m,$$

in terms of the fiber and matrix volume fractions. In terms of individual constituent properties, the mass fractions and volume fractions are related by:

$$W_f = \frac{\frac{\rho_f}{\rho_m}}{\frac{\rho_f}{\rho_m} V_f + V_m}$$
$$W_m = \frac{1}{\frac{\rho_f}{\rho_m}(1 - V_m) + V_m}$$

One should always state the basis of calculating the fiber content of a composite. It is given in terms of mass or volume. Based on above Equation, it is evident that volume and mass fractions are not equal and that the mismatch between the mass and volume fractions increases as the ratio between the density of fiber and matrix differs from one.

Density

The derivation of the density of the composite in terms of volume fractions is found as follows. The mass of composite w_c is the sum of the mass of the fibers w_f and the mass of the matrix w_m as:

$$w_c + w_f = w_m,$$

$$\rho_c v_c = \rho_f v_f + \rho_m v_m, \quad \rho_c = \rho_f \frac{v_f}{v_c} + \rho_m \frac{v_m}{v_c}$$

Using the definitions of fiber and matrix volume fractions from Equation:

$$\rho_c = \rho_f V_f + \rho_m V_m.$$

Now, consider that the volume of a composite v_c is the sum of the volumes of the fiber v_f and matrix (v_m):

$$v_c = v_f + v_m.$$

The density of the composite in terms of mass fractions can be found as:

$$\frac{1}{\rho_c} = \frac{W_f}{\rho_f} + \frac{W_m}{\rho_m}.$$

Table: Typical properties of fibers (SI system of units).

Property	Units	Graphite	Glass	Aramide
Axial modulus	GPa	230	85	124
Transverse modulus	GPa	22	85	8
Axial Poisson's ratio	_	0.30	0.20	0.36
Transverse Poisson's ratio	_	0.30	0.20	0.37
Axial shear modulus	GPa	22	35.42	3
Axial Coefficient of thermal expansion	μm/m/°C	-1.3	5	-5.0
Transverse Coefficient of thermal expansion	μm/m/°C	7.0	5	4.1
Axial tensile strength	MPa	2067	1550	1379
Axial compressive strength	MPa	1999	1550	276
Transverse tensile strength	MPa	77	1550	7
Transverse compressive strength	MPa	42	1550	7
Shear Strength	MPa	36	35	21
Specific gravity	-	1.8	2.5	1.4

Table: Typical properties of Matrices (SI system of units).

Property	Units	Epoxy	Aluminum	Polyamide
Axial modulus	GPa	3.4	71	3.5
Transverse modulus	GPa	3.4	71	3.5
Axial Poisson's ratio	_	0.30	0.30	0.35
Transverse Poisson's ratio	_	0.30	0.30	0.35
Axial shear modulus	GPa	1.308	27	1.3

Coefficient of thermal expansion	μm/m/°C	63	23	90
Coefficient of moisture expansion	m/m/kg/kg	0.33	0.00	0.33
Axial tensile strength	MPa	72	276	54
Axial compressive strength	MPa	102	276	108
Transverse tensile strength	MPa	72	276	54
Transverse compressive strength	MPa	102	276	108
Shear Strength	MPa	34	138	54
Specific gravity	-	1.2	2.7	1.2

Idealization of Microstructure of Fibrous Composite

The micromechanics is a study at fibre and matrix level. Thus, the geometry of arrangement of the fibres and matrix in a composite is an essential requirement to develop a model for the study. Some of the methods do not use the geometry of arrangement. Most of the methods developed for micromechanical analysis assume that:

- The fibers and matrix are perfectly bonded and there is no slip between them.
- The fibres are continuous and parallel.
- The fibres are assumed to be circular in cross section with a uniform diameter along its length.
- The space between the fibres is uniform throughout the composite.
- The elastic, thermal and hygral properties of fibre and matrix are known and uniform.
- The fibres and matrix obey Hooke's law.
- The fibres and the matrix are only two phases in the composite.
- There are no voids in the composite.

There are many ways to idealize the cross section of a lamina. In Figure 1 are shown two popular idealizations. The most commonly preferred arrangements are square packed and hexagonal packed arrays of fibres in matrix. The square and hexagonal packed arrays can be as shown in Figure (a), and (b), respectively.

In these idealizations it is seen that due to symmetry and periodicity of these arrays one can consider only one array to analyze the lamina at micro scale. Further, if this one array represents the general arrangement of fibres with respect to matrix and the interactions of fibre and matrix phases, then such array is called Representative Volume Element (RVE). Further, this RVE as a volume of material statistically represents a homogeneous material. In the analysis of an RVE the boundary conditions are chosen such that they reflect the periodicity. Thus, the arrays shown in figure are various RVEs. One should be able to see that the RVE also reflects the volume fractions. The term RVE was first coined by Hill in 1963. For example, the square RVE represents a lower fibre volume fraction than a hexagonal RVE. Note that RVE is also called as Unit Cell.

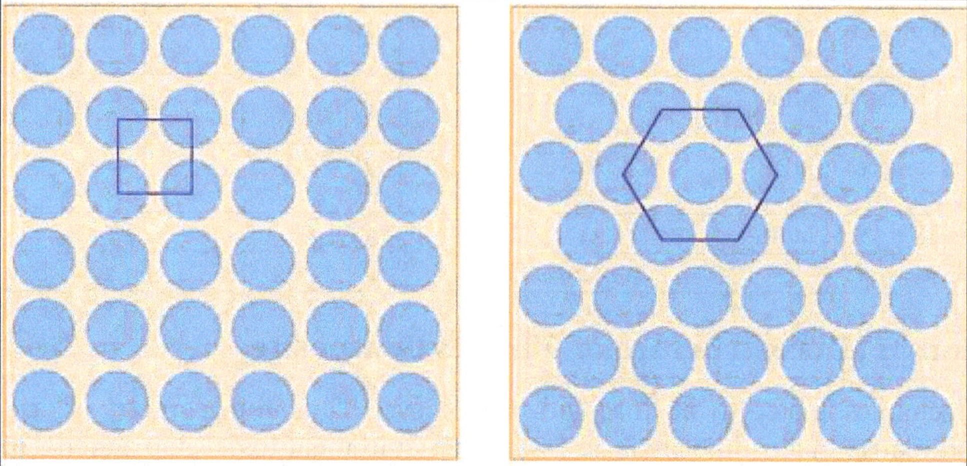

(a) Square packed array (b) Hexagonal packed array

Idealization of cross section of lamina.

Strength of Material Approximations

In general, the laminates made are thin. Hence, for such laminates the analysis done using Kirchhoff and plane stress assumptions is reasonably good. For such analysis, one needs the engineering constants that occur in defining planar constitutive equations. These engineering constants are:

- E_1^* : The axial modulus
- $E_2^* = E_3^*$: Transverse modulus
- $v_{12}^* = v_{13}^*$: Axial Poisson's ratio (for loading in x_1 - direction)
- $G_{12}^* = G_{13}^*$: Axial shear modulus (shear stress parallel to the fibers)

Further, it is seen that for transversely isotropic composite, four out of five (the fifth one is G_{23}^*) properties can be developed from this approach. For the planar hygro-thermal analysis of such laminates, one can also obtain the in-plane coefficients of thermal expansions α_1^* and α_2^* and hygroscopic expansion β_1^* and β_2^* as well.

It is important to note that this approach involves assumptions which do not necessarily satisfy the requirements of an exact elasticity solution. In this approach the effective properties will be expressed in terms of the elastic properties and volume fractions of the fiber and matrix. The lamina is considered to be an alternate arrangement of fibres and matrix. The RVE chosen in these derivations is shown in figure. The RVE here does not take into account the cross sectional arrangement of fibres and matrix, rather it represents volume of the material through the cross sectional area of fibre and matrix.

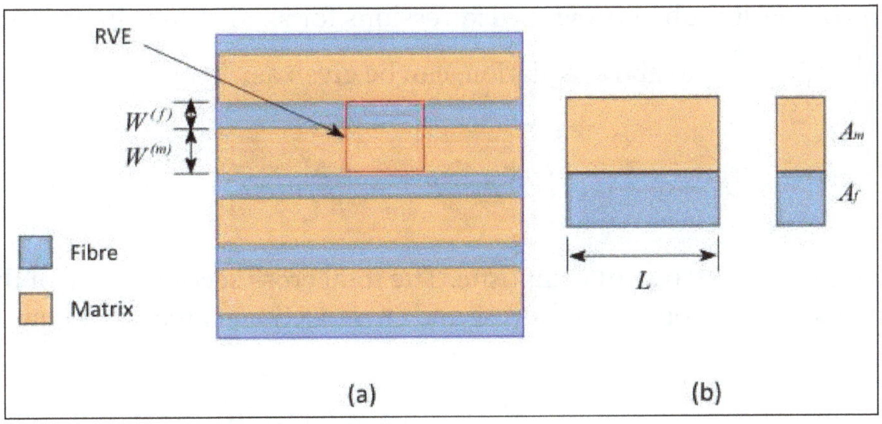

(a) Unidirectional lamina, (b) RVE for unidirectional composite for prediction of elastic properties.

Let, A_f and A_m represent fibre area and matrix area, respectively. $W^{(f)}$ and $W^{(m)}$ represent fibre and matrix widths, respectively. L be the length of the RVE.

Effective Axial Modulus E_1^*

The unit cell as shown in figure is used to compute the effective axial modulus E_1^*. It should be noted that the thickness of the unit cell is not important in this computation. Further, the cross sectional shapes are not considered in this calculation. However, the cross sectional areas are important in this calculation. The thicknesses of the fibre and matrix constituents are same in the unit cell. Hence, the areas of the constituents represent the volume fractions of the constituents.

In the calculation of effective axial modulus, it is assumed that the axial strain in the composite is uniform such that the axial strains in the fibers and matrix are identical. This assumption is justified by the fact that the fibre and the matrix in the unit cell are perfectly bonded. Hence, the elongation in the axial direction of the fibre and matrix will also be identical. Thus, the strains in the fibre and matrix can be given as:

$$\bar{\varepsilon}_1 = \varepsilon_1^{(f)} = \varepsilon_1^{(m)} = \frac{\Delta L}{L},$$

where, $\bar{\varepsilon}_1$ is the axial strain in the composite and $\varepsilon_1^{(f)}$ and $\varepsilon_1^{(m)}$ are the axial strains in fibre and matrix, respectively. Now, let $\varepsilon_1^{(f)}$ and $\varepsilon^{(m)}$ be the axial Young's moduli of the fibre and matrix, respectively. We can give the axial stress in the fibre, $\sigma_1^{(f)}$ and matrix, $\sigma_1^{(m)}$ as:

$$\sigma_1^{(f)} = E_1^{(f)} \varepsilon_1^{(f)} \quad \text{and} \quad \sigma_1^{(m)} = E^{(m)} \varepsilon_1^{(m)}.$$

Using the above equation and the cross section areas of the respective constituent in the unit cell, we can calculate the forces in them as:

$$F_1^{(f)} = \sigma_1^{(f)} A_f \quad \text{and} \quad F_1^{(m)} = \sigma_1^{(m)} A_m.$$

The total axial force in the composite is sum of the axial forces in fibre and matrix. Thus, the total axial force in the composite substituting the expressions for axial strains in fibre and matrix from Equation $\bar{\varepsilon}_1 = \varepsilon_1^{(f)} = \varepsilon_1^{(m)} = \frac{\Delta L}{L}$ in above equation, can be given as:

$$F_1 = F_1^{(f)} + F_1^{(m)} = \sigma_1^{(f)} A_f + \sigma_1^{(m)} A_m = \left(E_1^{(f)} A_f + E^{(m)} A_m\right) \frac{\Delta L}{L}$$

Now $\bar{\sigma}_1$ be the average axial stress in composite. The total cross sectional area of the composite is $A = A_f + A_m$. Thus, using the average axial stress and cross sectional area of the composite, the axial force is:

$$F_1 = \bar{\sigma}_1 A$$

Thus, combining above two Equations and rearranging, we get:

$$\bar{\sigma}_1 = \left(E_1^{(f)} \frac{A_f}{A} + E^{(m)} \frac{A_m}{A}\right) \frac{\Delta L}{L}$$

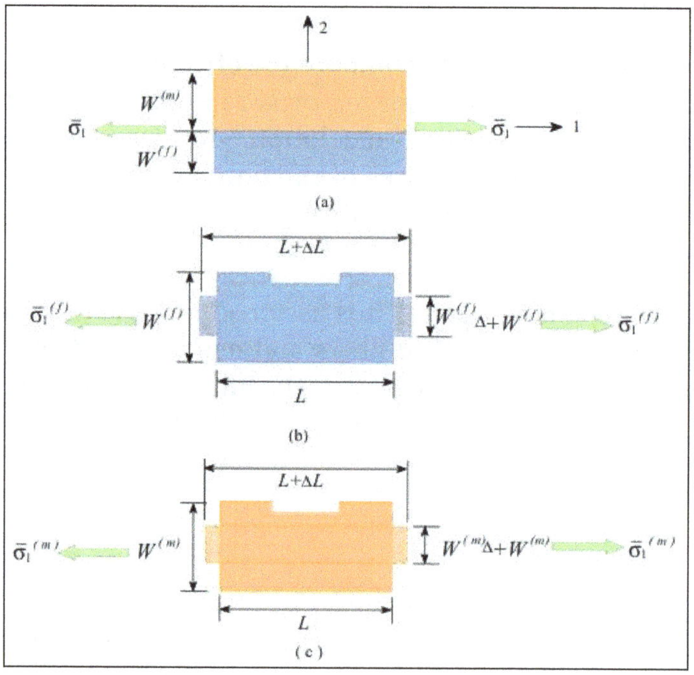

(a) Undeformed unit cell under $\bar{\sigma}_1$ (b) and (c) deformed individual constituents of the unit cell.

Let us define:

$$\bar{\sigma}_1 = E_1^* \bar{\varepsilon}_1 = E_1^* \frac{\Delta L}{L},$$

Further, noting that the ratios $\frac{A_f}{A}$ and $\frac{A_m}{A}$ for same length of fibre and matrix represent the fibre and matrix volume fractions, respectively. Thus, combining above two Equations, we get:

$$E_1^* = E_1^{(f)} V_f + E^{(m)} V_m = E_1^{(f)} V_f + E^{(m)} (1 - V_f).$$

The above equation relates the axial modulus of the composite to the axial moduli of the fibre and matrix through their volume fractions. Thus, the effective axial modulus is a linear function of the fiber volume fraction. This equation is known as rule of mixtures equation. It should be noted that the effective properties are functions of the fiber volume fractions; hence it should always be quoted in reporting the effective properties of a composite.

Effective Axial (Major) Poison's Ratio v_{12}^{*}

To determine the effective axial Poisson's ratio we consider the loading as in the case applied for determining the effective axial modulus. Here, for this loading we have $\bar{\sigma}_1 \neq 0$ and other stresses are zero. We define the effective axial Poisson's ratio as:

$$v_{12}^{*} = -\frac{\bar{\varepsilon}_2}{\bar{\varepsilon}_1}.$$

The effective strain in direction 2 from figure above can be given as:

$$\bar{\varepsilon}_2 = \frac{\Delta W}{W} = \frac{\Delta W^{(f)} + \Delta W^{(m)}}{W^{(f)} + W^{(m)}},$$

Now, the changes in $W^{(f)}$ and $W^{(m)}$ can be obtained using the Poisson's ratio of individual constituents. The axial Poisson's ratios for fibre and matrix are given as:

$$v_{12}^{(f)} = -\frac{\varepsilon_2^{(f)}}{\varepsilon_1^{(f)}} = -\frac{\Delta W^{(f)} / W^{(f)}}{\Delta L / L} \quad \text{and} \quad v^{(m)} = \frac{\varepsilon_2^{(m)}}{\varepsilon_1^{(m)}} = -\frac{\Delta W^{(m)} / W^{(m)}}{\Delta L / L}.$$

Thus, the changes in $W^{(f)}$ and $W^{(m)}$ are given as:

$$\Delta W^{(f)} = -v_{12}^{(f)} W^{(f)} \frac{\Delta L}{L} \quad \text{and} \quad \Delta W^{(m)} = -v^{(m)} W^{(m)} \frac{\Delta L}{L},$$

The total change in W is given as:

$$\Delta W = \Delta W^{(f)} + \Delta W^{(m)}.$$

The strain in direction 2 for the composite can be given using Equation $E_1^{*} = E_1^{(f)} V_f + E^{(m)} V_m = E_1^{(f)} V_f + E^{(m)} (1 - V_f)$ and Equation $\Delta W = \Delta W^{(f)} + \Delta W^{(m)}$ as:

$$\bar{\varepsilon}_2 = \frac{\Delta W}{W} = \frac{\Delta W^{(f)} + \Delta W^{(m)}}{W} = -\left(v_{12}^{(f)} \frac{W^{(f)}}{W} + v^{(m)} \frac{W^{(m)}}{W} \right) \frac{\Delta L}{L},$$

Here, $\frac{W^{(f)}}{W}$ and $\frac{W^{(m)}}{W}$ denote the fibre and matrix volume fractions for same length of fibre and matrix. Note that $\frac{\Delta L}{L}$ denotes the effective axial strain $\bar{\varepsilon}_1$. Thus, from Equation above the effective axial Poisson's ratio is written as:

$$v_{12}^{(*)} = v_{12}^{(f)} + V_f + v^{(m)} V_m .$$

The above equation is the rule of mixtures expression for composite axial Poisson's ratio.

Effective Transverse Modulus E_2^*

Here, we are going to derive the effective transverse modulus by loading the RVE in direction 2 as shown in Figure below. There are two considerations while deriving this effective modulus. The first approach considers that the deformation of the each constituent is independent of each other as shown in Figure bwlow and the deformation in direction 1 is not considered. The second approach considers that deformations of the fibre and matrix in direction 1 are identical as they are perfectly bonded. To calculate the effective modulus in direction 2, a stress $\bar{\sigma}_2$ is applied to the RVE as shown in figure.

First Approach

As mentioned, the fibre and matrix deform independently of each other. The resulting deformation in direction 1 is not considered here. This assumption is simplistic and was used by early researchers. The fibre and matrix are subjected to same state of stress. The state of stress is unidirectional, that is, $\bar{\sigma}_2^{(f)} = \bar{\sigma}_2^{(m)} = \bar{\sigma}_2$. Now, using the individual moduli and deformations in direction 2, these stresses can be given as:

$$\bar{\sigma}_2^{(f)} = E_2^{(f)} \varepsilon_2^{(f)} = E_2^{(f)} \frac{\Delta W^{(f)}}{W^{(f)}}$$

$$\bar{\sigma}_2^{(m)} = E^{(m)} \varepsilon_2^{(m)} = E^{(m)} \frac{\Delta W^{(m)}}{W^{(m)}},$$

From this equation we can write the individual deformations, which give the total deformation in direction 2 as:

$$\Delta W = \Delta W^{(f)} + \Delta W^{(m)} = \left(\frac{W^{(f)}}{E_2^{(f)}} + \frac{W^{(m)}}{E^{(m)}} \right) \bar{\sigma}_2,$$

Now, the composite strain in direction 2 can be calculated from the definition as:

$$\varepsilon_2 = \frac{\Delta W}{W} = \left(\frac{W^{(f)}}{W} \frac{1}{E_2^{(f)}} + \frac{W^{(m)}}{W} \frac{1}{E^{(m)}} \right) \bar{\sigma}_2 .$$

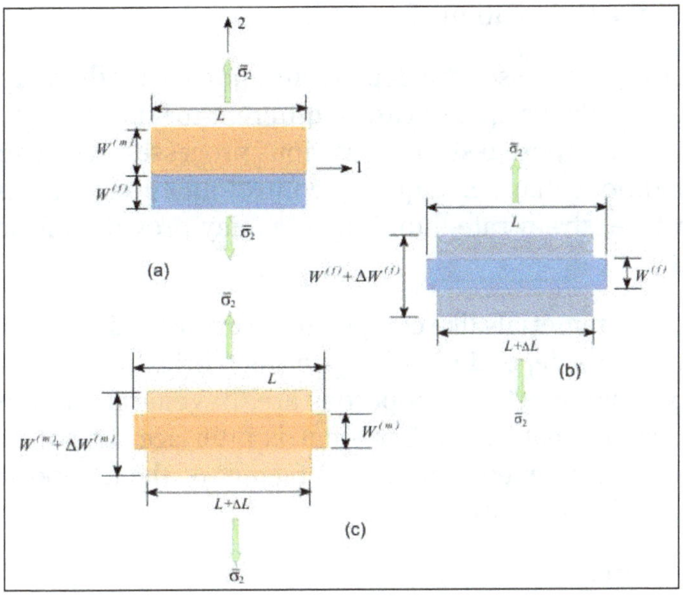

(a) Undeformed unit cell under uniform $\overline{\sigma}_2$ stress (b) and (c) deformed individual constituents of the unit cell.

Introducing the volume fractions in the above equation:

$$\overline{\varepsilon}_2 = \left(V_f \frac{1}{E_2^{(f)}} + V_m \frac{1}{E^{(m)}} \right) \overline{\sigma}_2 ,$$

Noting that $\frac{\overline{\sigma}_2}{\overline{\varepsilon}_2} = E_2^*$, from the above equation, we get:

$$\frac{1}{E_2^*} = \frac{V_f}{E_2^{(f)}} + \frac{V_m}{E^{(m)}} = \frac{V_f}{E_2^{(f)}} + \frac{(1 - V_f)}{E^{(m)}} .$$

This equation is the rule of mixtures equation for effective modulus E_2^*.

Mechanical Testing of Composites

The development of the mechanical testing of the materials depends upon other scientific factors. These factors help in better understanding and facilitate the progress in evaluating the various processes. These processes include:

- Quality control of a process.
- Quality assurance for the material developed and structure fabricated from thereof.
- Better material selection.
- Comparisons between available materials.
- Can be used as indicators in materials development programmes.
- Design analysis.
- Predictions of performance under conditions other than test conditions.

- Starting points in the formulation of new theories.

It should be noted that these processes are dependent upon each other. However, if they are considered individually then the data required can be different for the evaluation. For example, some tests are carried out as multipurpose tests using various processes. A conventional tensile test carried out under fixed conditions may serve quality control function whereas one carried out varying factors like temperature, strain rate, humidity etc. may provide information on load bearing capacity of the material.

The properties evaluated for materials like composite is very sensitive to various internal structure factors. However, these factors depend mainly upon the fabrication process or other factors. The internal structure factors that affect the properties are, in general, at atomic or molecular level. These factors mostly affect the matrix and fibre-matrix interface structure. The mechanical properties of the fibrous composite depend on several factors of the composition. These factors are listed below again for the sake of completeness.

- Properties of the fibre.
- Surface character of the fibre.
- Properties of the matrix material.
- Properties of any other phase.
- Volume fraction of the second phase (and of any other phase).
- Spatial distribution and alignment of the second phase (including fabric weave).
- Nature of the interfaces.

Another important factor is processing of the composites. There are many parameters that control the processing of composites that access the quality of adhesion between fibre and matrix, physical integrity and the overall quality of the final structure.

In case of composite the spatial distribution and alignment of fibres are the most dominating factor which causes the variation of properties. The spatial distribution and alignment of the fibres can change during the same fabrication process. Thus, for a given fabrication process the property evaluated from the composite material may show a large variation.

Tensile Testing

The well-known purpose of the tensile testing is to measure the ultimate tensile strength and modulus of the composite. However, one can measure the axial Poison's ratio with additional instrumentations. The standard specimen used for tensile testing of continuous fiber composites is a flat, straight-sided coupon. A flat coupons in ASTM standard D 3039/D 3039M-93 for 0° and 90° have been shown.

The specimen, as mentioned above is flat rectangular coupon. The tabs are recommended for gripping the specimen. It protects the specimen from load being directly applied to the specimen causing the damage. Thus, the load is applied to the specimen through the grips. Further, it protects the outer fibres of the materials. The tabs can be fabricated from a variety of materials, including fiberglass, copper, aluminum or the material and laminate being tested. When the tabs of composite material are used then according to ASTM specifications the inner plies of the tabs should match with the outer plies of the composite. This avoids the unwanted shear stresses at the interface of the specimen and tabs. However, the recent versions of the ASTM standards allow the use of tabs with reinforcement at ±45°. Further, end-tabs can also facilitate accurate alignment of the specimen in the test machine, provided that they are symmetrical and properly positioned on the specimen. The tabs are pasted to the specimen firmly with adhesive.

This specimen can provide data on:

- The axial modulus E_x.
- In-plane and through thickness Poisson's ratio γ_{xy}, γ_{xz}.
- Tensile ultimate stress σ_x^{ult}.
- Tensile ultimate strain ε_x^{ult}.
- Any nonlinear, inelastic response.

In general, the tensile tests are done on coupons with 0° laminae/laminate for corresponding axial properties and coupons with 90° laminae/laminate for corresponding transverse properties. The off axis laminae specimen also provides data on coefficient of mutual influence and the in-plane shear response.

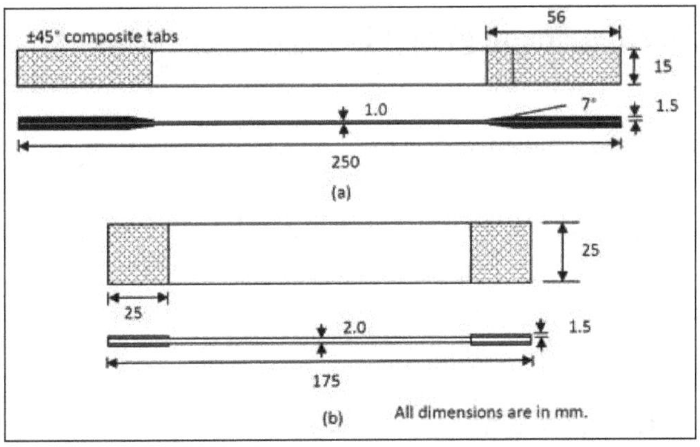

Composite tensile test specimens: (a) ASTM D 3039 for 0° and (b) ASTM D 3039 for 90°.

For orthotropic, symmetric laminates with 0° and 90° laminae, the effective axial modulus and Poisson's ratio is given as:

$$E_x = \frac{\sigma_{xx}}{\varepsilon_{xx}} = \frac{N_{xx}}{t\varepsilon_{xx}}, v_{xy} = -\frac{\varepsilon_{yy}}{\varepsilon_{xx}}.$$

The tensile strength is defined as the average stress at failure. Thus, the tensile strength can be given using the maximum applied force per unit length N_x and thickness t as:

$$\sigma_x^{-ult} = \frac{N_x^{max}}{t},$$

It should be noted that the failure of laminates is often influenced by inter laminar stresses along the free edge effects of the coupon. These factors will be explained in brief in one of the lecture. The measurement of tensile strength by experiments can also provide information on the comparison of laminate theory with experiments.

Off-Axis Laminae

One can measure the tensile properties by conducting experiments on off-axis laminae. However, there are certain issues associated with this kind of experiments. For example, the presence of axial shear coupling is associated with the nonzero α^*_{16}. Alternately, one can say that this term is associated with coefficient of mutual influence $\eta_{xy,x}$. Hence, these tests are not straight forward as in case of symmetric laminates with 0° and 90° laminae. Therefore, sometimes these tests are called as specialized tests.

When the experiments are conducted to measure the properties like E_x, v_{xy} and v_{xz} one can get the other properties along with these tests. For example, the coefficient of mutual influence $\eta_{xy,x}$, the nonlinear response and strength of an off-axis lamina for given fibre orientation can also be obtained.

There is an important issue associated with these tests is that what boundary conditions one should impose on the specimen? If a pure, uniform state of axial stress can be applied to the ends and sides of a specimen and the specimen is free to assume any desired deformation pattern, the state of stress will be uniform and constant through-out the specimen.

For uniform, far-field axial stress loading, that is $\sigma_{xx} \neq 0$, the stresses in principal material directions can be given as:

$$\sigma_{11} = m^2\sigma_{xx}, \sigma_{22} = n^2\sigma_{xx}, \tau_{12} = -mn\sigma_{xx}.$$

Further, the global elastic constants associated with axial stress loading are measured as:

$$E_x = \frac{\sigma_{xx}}{\varepsilon_{xx}}, v_{xy} = -\frac{\varepsilon_{yy}}{\varepsilon_{xx}}, \eta_{xy,x} = \frac{\gamma_{xy}}{\varepsilon_{xx}}.$$

Combining above two equations, we get:

$$\varepsilon_{11} = \frac{\sigma_{11}}{E_1} - \frac{v_{21}\sigma_{22}}{E_2} = \left(m^2 - n^2 v_{12}\right)\frac{\sigma_{xx}}{E_1},$$

$$\varepsilon_{22} = \frac{v_{12}\sigma_{11}}{E_1} + \frac{\sigma_{22}}{E_2} = \left(-\frac{v_{12}m^2}{E_1} + \frac{n^2}{E_2}\right)\sigma_{xx}$$

$$\gamma_{12} = \frac{\tau_{12}}{G_{12}} = \frac{-mn\ \sigma_{xx}}{G_{12}},$$

From the above equation all three strain components can be obtained for non-zero value of axial stress. Thus, from the third of the above equation we can find the shear modulus.

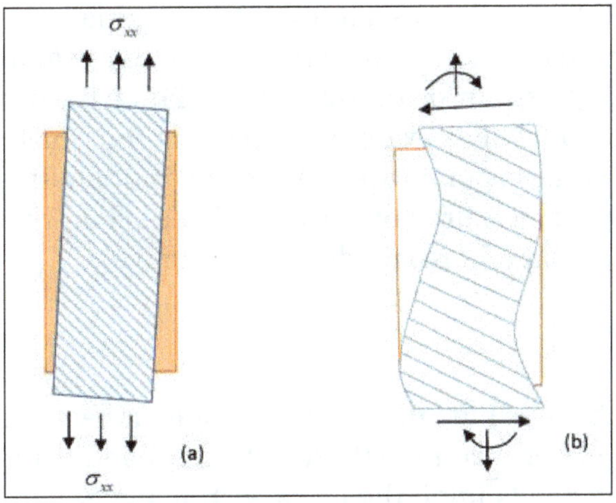

Axial load on off-axis laminae with effect of end constraint: (a) unconstrained displacement and (b) constrained displacement.

It is not easy to apply pure, uniform tensile stress to an off-axis coupon. The specimens are gripped in such a manner that the ends of the specimen are constrained and boundary condition is actually a specification of the axial end displacement. Further, there are more issues with these tests like the constrained displacement induces a doubly curved displacement field in the specimen. The deformed shape of the coupon with restrictions on the ends is depicted in Figure above. We will not deal the complete analysis for the measurements of the properties with tests on off-axis laminae.

The bone shaped specimens for chopper-fiber, metal matrix composite tensile tests. More details can be seen in ASTM D3552-77. Further, for the tensile testing for transverse properties of hoop wound polymer matrix composite cylinders are used. The details of this testing can be seen in ASTM D5450/D5450M-93.

Measurement of Modulus

It should be noted that due to progressive damage the stiffness of the lamina or laminae/laminate changes causing the stress strain curve to be non-linear. The measurement of modulus in a tensile testing from a non-linear loading curve can be done by three methods.

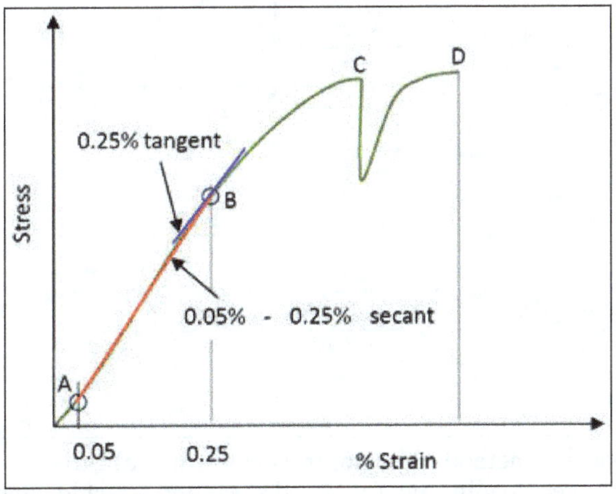

Typical tensile stress-strain curve with details.

In the first method the modulus is taken as a tangent to the initial part of the curve. In the second method a tangent is constructed at a specified strain level. For example, in the Figure above the modulus is measured at 0.25% strain or 0.0025 strain (Point B). In the third method, a secant is constructed between two points. For example a secant is constructed between points A and B. Typically, the strain values at these points are 0.0005 and 0.0025. In ASTM standards the secant is called as chord. The modulus measured by these methods is known as 'initial tangent modulus', 'B% modulus' and 'A%-B% secant (chord) modulus', respectively.

Compression Testing

Most of the structural members include the compression members. Such members can be loaded directly in compression or under a combination of flexural and compression loading. The axial stiffness of such members depends upon the cross-sectional area. Thus, it is proportional to the weight of the structure. One can alter the stiffness by changing the geometry of the cross section within limits. However, some of the composites have low compressive strength and this fact limits the full potential application of these composites.

The compression testing of the composites is very challenging due to various reasons. The application of compressive load on the cross section can be done in three ways: directly apply the compressive load on the ends of a specimen, loading the edges in shear and mixed shear and direct loading. These three ways of imposing the loads for compression testing are shown.

During compression loading the buckling of the specimen should be avoided. This demands a special requirement on the holding of the specimen for loading purpose. Further, it demands for special geometry of the specimen. These specimens are smaller in size as compared to the tensile testing specimens. A compression test specimen according to ASTM D695 (modified) standard is shown.

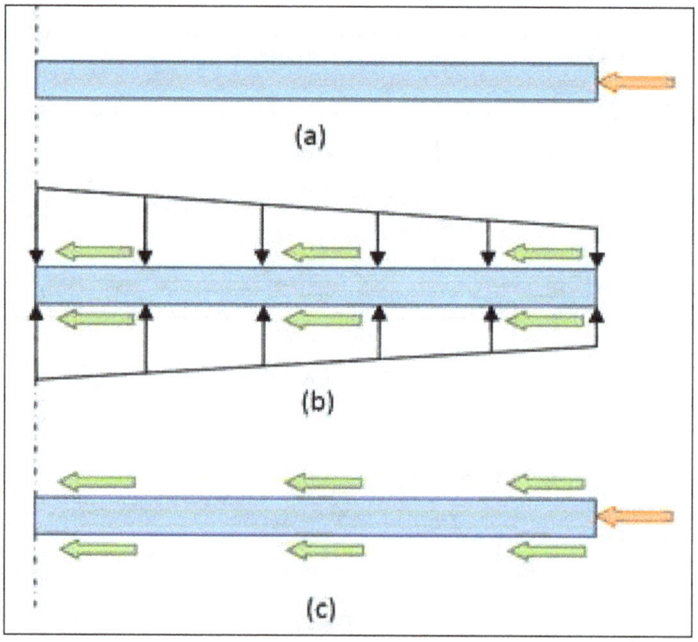

Load imposition methods for compression testing. (a) Direct end loading (b) Shear loading and (c) Mixed shear and direct loading.

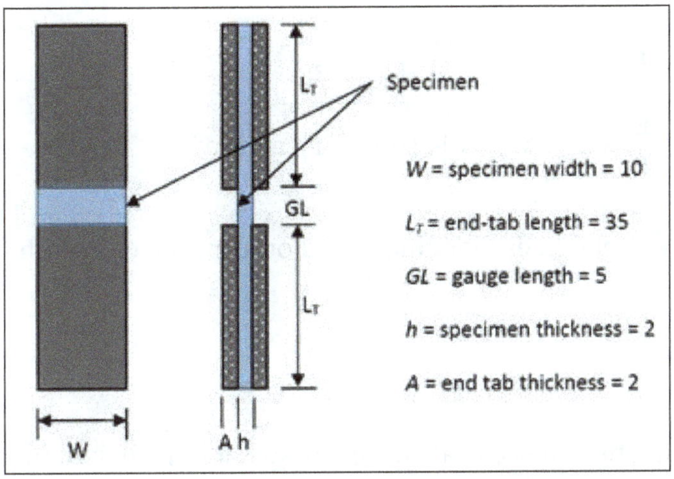

Composite compression test specimen according to ASTM D695 (modified) standard.

Shear Testing

Here we will see measurement of in-plane shear modulus G_{12} only. The methods are listed below:

- Tension of a $[\pm 45]_s$ laminate.
- Tension of an off-axis lamina.
- Torsion of a unidirectional tube.
- Iosipescu shear of unidirectional laminae and cross ply laminates.
- Rail shear of unidirectional laminae.
- Picture frame test.

$[\pm 45]_s$ Tensile Test

A tension test on $[\pm 45]_s$ laminate is popularly used test for the measurement of in-plane shear modulus G_{12}. The more details of this test are available in ATSM standard D3518/D3518/M–91. According to ASTM standard the method uses a 250 mm long rectangular specimen with width 25 mm and thickness 2 mm. Further, it is recommended that for materials constructed with layers thicker than 0.125mm, the laminate should consist of 16 layers, that is, $[\pm 45]_s$. The specimen is shown. The dimensions in this figure are in mm.

When a $[\pm 45]_{4s}$ is subjected to axial tensile stress $\bar{\sigma}_{xx}$ then stresses in principal material coordinates developed in each of the +45° and -45° lamina are given as:

$$\sigma_{11} = B\bar{\sigma}_{xx}$$
$$\sigma_{22} = (1-B)\bar{\sigma}_{xx}$$
$$\tau_{12} = \frac{-1}{2mn}\left[B\left(1-2m^2\right)+m^2\right]\bar{\sigma}_{xx}$$

where,

$$B = \left[\frac{m^2\left(2m^2-1\right)+4m^2n^2\dfrac{G_{12}}{E_2}\left(\dfrac{E_2}{E_1}v_{12}+1\right)}{4m^2n^2\dfrac{G_{12}}{E_2}\left(\dfrac{E_2}{E_1}+2\dfrac{E_2}{E_1}v_{12}+1\right)+\left(2m^2-1\right)\left(m^2-n^2\right)} \right]$$

and other quantities as defined in earlier chapters. For a special case with $\theta = 45°$ we get the shear stress as:

$$\tau_{12} = (\pm\theta) = \mp\frac{\bar{\sigma}_{xx}}{2}$$

Thus, from this equation one can see that the shear stress in principal material directions is statically determinate, that is, it is independent of material properties of the specimen and only depends upon the magnitude of the applied stress. The magnitude of this stress is half of the applied stress.

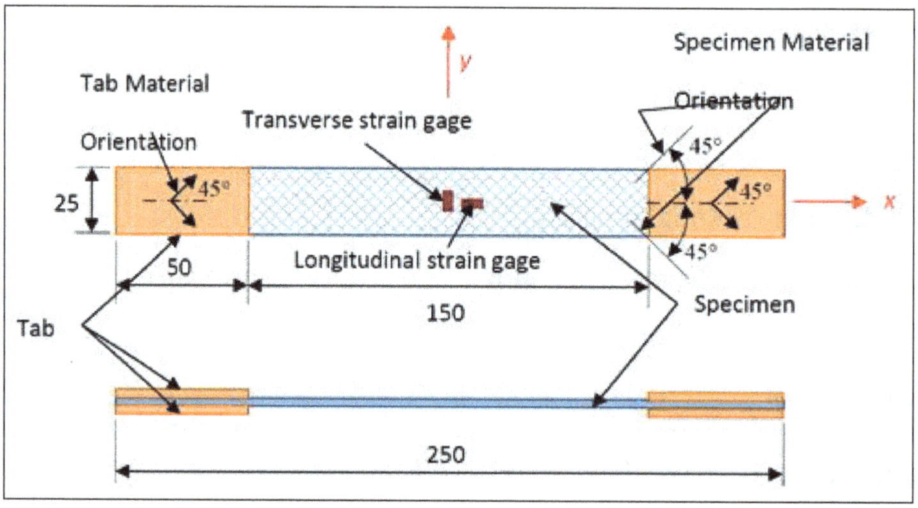

Specimen geometry and strain gage positioning for $[\pm 45]_{4s}$ tensile testing.

From the knowledge of linear elastic behaviour of the orthotropic materials it is clear that the shear response is uncoupled from the normal response. Hence, in-plane shear modulus G_{12} can be determined directly from a tensile test on a $[\pm 45]_{4s}$ laminate.

Now the shear strain γ_{12} in principal material coordinates can be found by transformation of the measured axial and transverse strains ε_{xx} and ε_{yy}. It should be noted that the shear strain γ_{xy} is zero for orthotropic laminates under tension γ_{12} and is independent of γ_{xy} for $\theta = \pm 45°$. Thus, from the strain transformation relations, we can get the shear strain in principal material directions as:

$$\gamma_{12} = -\left(\varepsilon_{xx} - \varepsilon_{yy}\right)$$

Thus, from the definition of the shear modulus we get:

$$G_{12} = \frac{\bar{\sigma}_{xx}}{2\left(\varepsilon_{xx} - \varepsilon_{yy}\right)}$$

The above equation can be rearranged in the following manner to express the shear modulus in terms of effective properties of $[\pm 45]_{4s}$ laminate.

$$G_{12} = \frac{\dfrac{\bar{\sigma}_{xx}}{\varepsilon_{xx}}}{2\left(\dfrac{\varepsilon_{xx}}{\varepsilon_{xx}} - \dfrac{\varepsilon_{yy}}{\varepsilon_{xx}}\right)} = \frac{E_x}{2\left(1+\nu_{xy}\right)},$$

Here, E_x is the effective modulus of the $[\pm 45]_{4s}$ laminate.

The measurement of in-plane shear modulus from shear stress-strain curve is done as follows. The shear stress-strain curve for ±45° specimen is obtained first. A typical shear stress-strain curve for such a specimen is shown. The shear modulus is obtained from the initial slope of the this curve in the range of 0.1-0.5% strain as:

$$G_{12} = \frac{\tau''_{12} - \tau'_{12}}{\gamma''_{12} - \gamma'_{12}}.$$

The tensile test on ±45° specimen provides an acceptable method for the measurement of in-plane shear modulus. However, one should be careful while interpreting the ultimate shear strength and strain. It should be noted that the laminae are subjected to a biaxial state of stress and not a pure shear. The normal stresses act along the shear planes causing the onset of mixed mode fracture. Other kind of failure like multiple ply cracking, fibre rotation and edge or internal delaminations occur prior to final failure. Therefore, the true failure is very difficult to determine. The shear strength is specified by different standards corresponding either to the ultimate load generated during the test or to a specified strain level. It is recommended in ISO standard that the test be terminated at $\gamma_{12} = 5\%$. The shear strength is taken as the peak load at or before 5% strain.

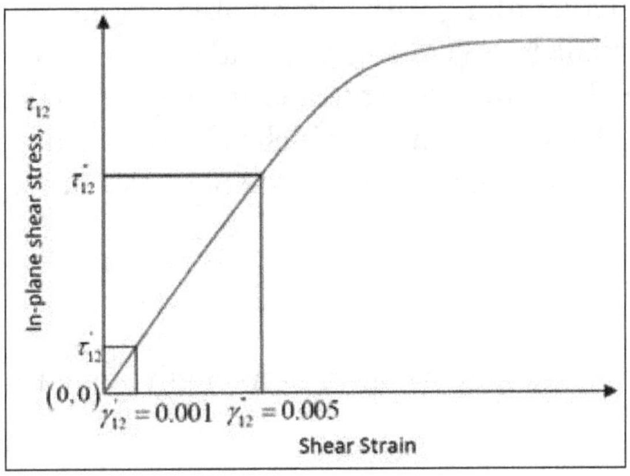

Typical shear stress-strain curve for specimen.

Shear of an Off-Axis Lamina

In similar way to the tensile testing of a $[\pm 45]_{4s}$ laminate one can use a unidirectional off-axis tensile coupon to determine the shear response of a composite in the principal material coordinates. A tensile test on 10° off-axis lamina is a commonly used. Specimen has same geometry. The state of stress in principal material coordinate directions can be obtained from transformation relations. Since, the shear response in the principal material coordinates is uncoupled from the normal response we can write the shear modulus as:

$$G_{12} = \frac{\tau_{12}}{\gamma_{12}},$$

The shear stress in the principal material directions due to axial tensile stress can be given using transformation relations as:

$$\bar{\tau}_{12} = -mn\bar{\sigma}_{xx},$$

The shear strain is measured from the strains $\varepsilon_{xx}, \varepsilon_{yy}$ and γ_{xy} with the help of strain transformation relations. Then the apparent shear modulus $\bar{G}_{12}$ can be given as:

$$\bar{G}_{12} = \frac{-mn\bar{\sigma}_{xx}}{\gamma_{12}}.$$

Rail Shear Test

This is a very popular method used to measure in-plane shear properties. This method is extensively used in aerospace industry. The shear loads are imposed on the edges of the laminate using specialized fixtures. There are two types of such fixtures: Two rail and three rail fixture. The ASTM D4255 standard covers the specification for two and three rail specimens for both continuous and discontinuous (0° and 90° fibre alignment), symmetric laminates and randomly oriented fibrous laminates.

Two Rail Shear

Test The two rail shear test fixture along with a laminate to be tested is shown. The Figure shows the specimen geometry according to ASTM D4255 standard. The two rail shear test fixture has two rigid parallel steel rails for loading purpose. The rails are aligned to the loading direction as shown. Thus, it induces the shear load in the specimen which is bolted to these rails. A strain gage is bonded at 45° to the longitudinal axis of the specimen. The Shear strength is obtained as:

$$\tau_{xy}^{ult} = \frac{P_{max}}{Lh},$$

where, p_{max} is ultimate failure load, L is the specimen length along the rails and h is the specimen thickness.

The shear modulus is given as:

$$G_{xy} = \frac{\Delta\tau_{xy}}{\Delta\gamma_{xy}} = \frac{\Delta P}{2Lh\Delta\varepsilon_{45}}.$$

Where, ΔP is the change in applied load and $\Delta\varepsilon_{45}$ is the change in strain for +45° or -45° strain gage in the initial linear stress-strain regime. It is suggested that the change in the strain is taken as the average of the change in strains on the both sides of the specimen. Various modes of failure are seen. The modes are highly dependent upon the microstructure of the material.

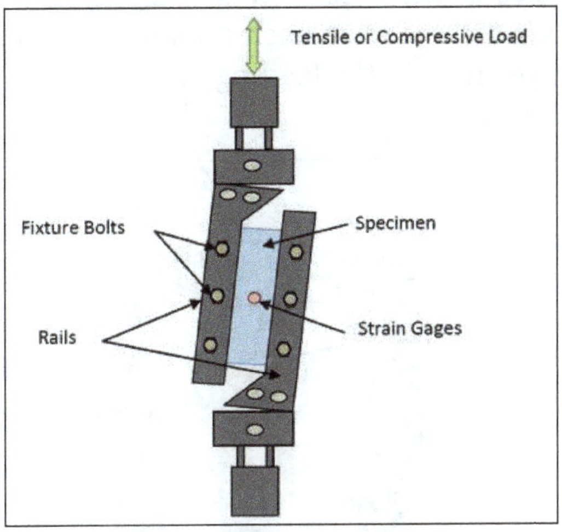

Two rail shear fixture for shear testing.

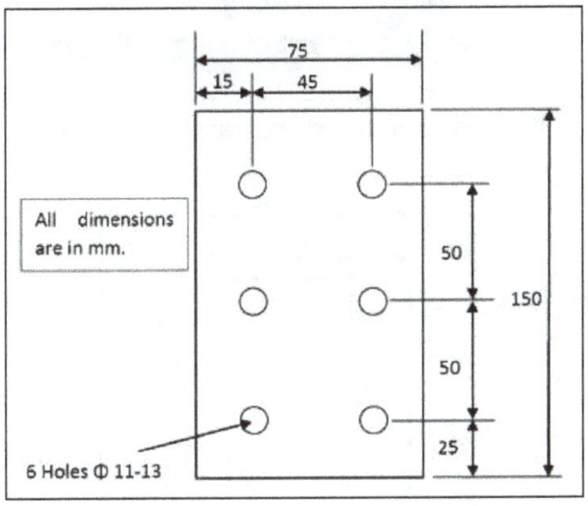

Two rail shear test specimen.

Three Rail (Symmetric) Shear Test

The three rail shear test is an improved version of the rail shear test. Using one more rail in two rail shear test fixture it can produce a closer approximation to pure shear. The fixture consists of 3 pairs of rails clamped to the test specimen as shown. The outside pairs are attached to a base plate which rests on the test machine. Another pair (third middle) pair of rails is guided through a slot in the top of the base fixture. The middle pair loaded in compression. The shear force in laminate is generated via friction between rail and specimen. The strain gages bonded to the specimen at 45° to the specimen's longitudinal axis. The specimen geometry is shown. The shear strength is given as:

$$\tau_{xy}^{ult} = \frac{P_{max}}{2Lh},$$

And the shear modulus is given as:

$$G_{xy} = \frac{\Delta\tau_{xy}}{\Delta\gamma_{xy}} = \frac{\Delta P}{4Lh\Delta\varepsilon_{45}},$$

Where, all variables in these two equations are given previously.

It should be noted that the holes in the specimen are slightly oversized than the bolts used for clamping. Further, the bolts are tightened in such a manner to ensure that there is no bearing contact between the bolt and specimen in the loading direction. It is recommended that each bolt is tightened with a 100 Nm torque.

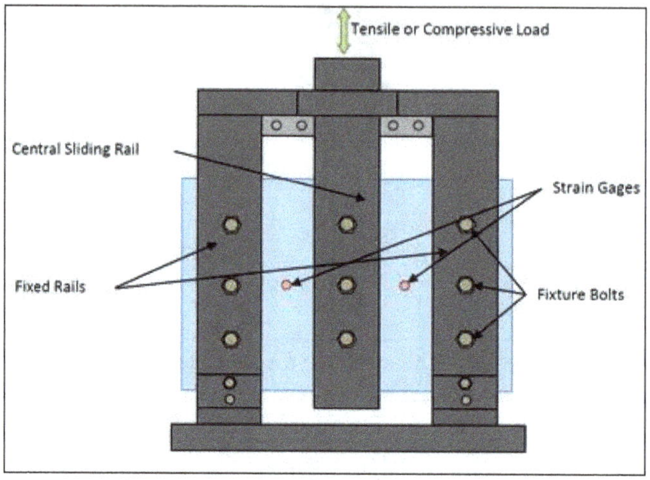

Three rail shear fixture for shear testing.

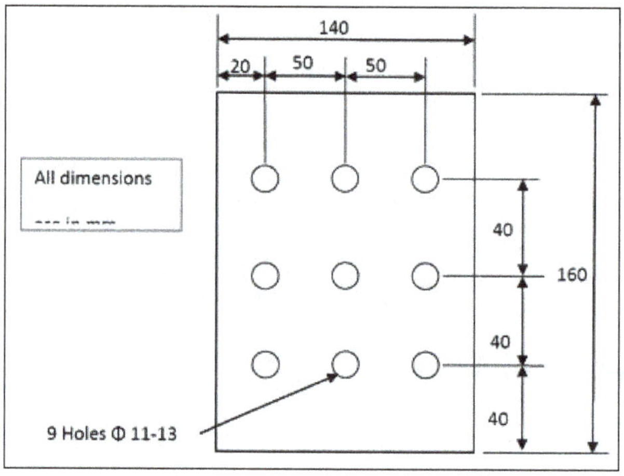

Specimen dimensions for three-rail shear test.

Flexural Tests

The flexural tests are conducted to determine the mechanical properties of resin and laminated fiber composite materials. Further, these tests are used to determine the inter laminar shear strength of a laminate, shear modulus, shear strength, tensile and compression moduli along with flexural and shear stiffness. These tests are not only used for composites but also for sandwich beams.

These tests are simple one. Further, they need simple instrumentation and equipment required. These tests conducted on beams of uniform cross section. These beam specimens do not require the end tabs. There are two methods to carry out these tests. The beam is a flat rectangular specimen and is simply supported close to its ends. In the first method the beam is centrally loaded. Thus gives three point bending. Since there are three important points (two end supports and one central loading point) along the span of the beam this method is called as three-point bending test.

In the second method the beam is loaded by two loads placed symmetrically between the supports. In this method there are four important points (two end supports and two loading points) along the span of the beam. Thus, it gives four-point bending. Hence, this method is called four point bending. These methods are shown schematically. Also shown in this figures are the shear force diagram (SFD) and bending moment diagrams (BMD) related to the particular loading regimes.

From the shear force and bending moment diagrams it is clear that there is a stress concentration at the point of loading. However, for four point bending there is uniform bending moment and both shear force and interlaminar shear stress are zero between the loading points. Thus, it leads to the pure bending loading. Such a state of stress is desirable in testing.

The properties are assumed to be uniform through the thickness as composite as it is a unidirectional composite or isotropic material. For such a material the normal stress varies linearly across the thickness. The maximum in compression is on one side and an equal maximum in tension on other side of the thickness and passes through zero at the mid-plane. The maximum normal stress is given as:

$$|\sigma_c| = |\sigma_T| = \frac{6M}{bh^2},$$

Where, M is the bending moment, b is width and h is the thickness of the specimen. Further, σ_c and σ_T denote compressive and tensile normal stresses, respectively. The shear stress varies parabolic through the thickness with maximum at mid plane and zero at the outer surface. The maximum shear stress at the mid plane is given as:

$$\tau = \frac{3Fs}{2bh},$$

Where Fs is the shear force on the specimen cross section. The normal stress and shear force variation through the thickness is shown. The flexural response of the beam in three or four point bending test is obtained by recording the load applied and the resulting strain. The resulting strains are measured using the strain gages bonded on the beam in the gage length. It is clear from the distribution of the shear force and bending moment that the state of stress in specimens subjected to three and four-point bending tests are somewhat different. Thus, it may lead to differences in the results.

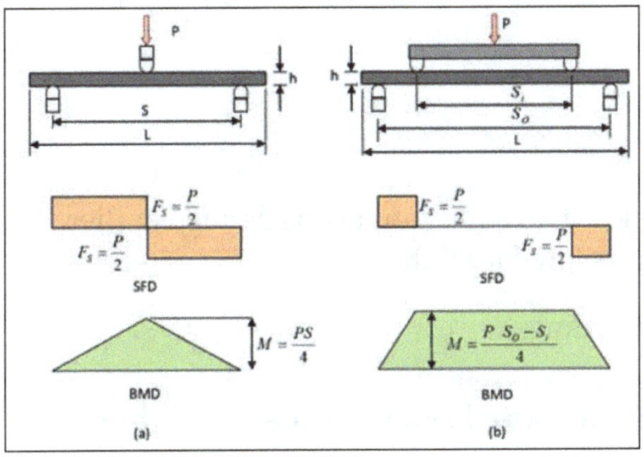

Shear force and bending moment diagrams for (a) three point and (b) four point bending test.

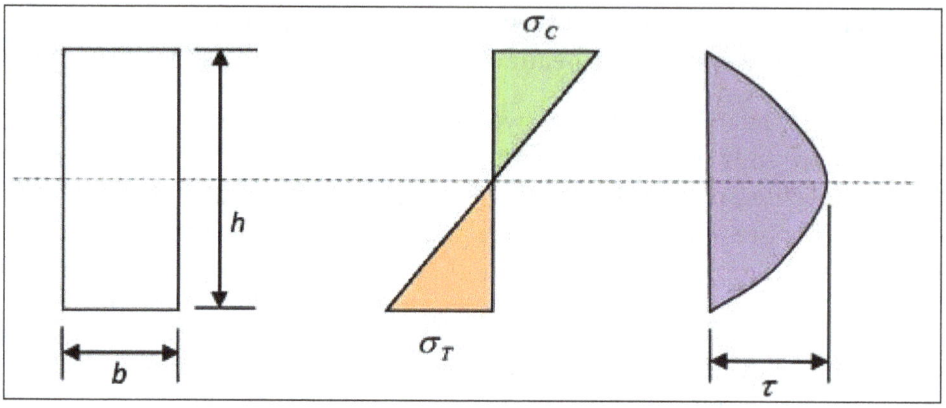

Bending and shearing stresses in the thickness direction.

In the following we will see the measurement of flexural modulus and maximum stress on the outer surface of the beam.

Flexural Strength

This is the stress on the surface of the specimen at failure, which should be accompanied by the breaking of fibers, rather than inter laminar shear. In the three point bending method the flexural modulus E_f is given as:

$$E_f = \frac{S^3 m}{4bh^3},$$

Where, E_f is flexural modulus, S is the support span, m is the slope of the load-deflection curve, b and h are the width and thickness of the specimen, respectively.

In case of four point bending there are two options according to ASTM D790 standard. In the first option the loading span is one third of the support span. For this case the flexural modulus is given as:

$$F_f = 0.21\frac{S^3 m}{bh^3}.$$

In the second option the loading span is half of the support span. The flexural modulus for this case is given as:

$$F_f = 0.17\frac{S^3 m}{bh^3},$$

Where, the parameters in these two equations are as defined earlier. The maximum stress on outer surface of the beam is given below for all the cases.

- $\sigma = \frac{3PS}{2bh^2}$: 3 point bending.
- $\sigma = \frac{PS}{bh^2}$: 4 point bending with loading span equal to third support span.

- $\sigma = \dfrac{3PS}{4bh^2}$: 4 point bending with loading span equal to one third support span.

It is important to note that the measurement of width and thickness of the beam is important for accurate measurement of flexural modulus and maximum stresses.

Mechanical Properties of Laminates

- The laminate consists of perfectly bonded layers. There is no slip between the adjacent layers. In other words, it is equivalent to saying that the displacement components are continuous through the thickness.
- Each lamina is considered to be a homogeneous layer such that its effective properties are known.
- Each lamina is in a state of plane stress.
- The individual lamina can be isotropic, orthotropic or transversely isotropic.
- The laminate deforms according to the Kirchhoff - Love assumptions for bending and stretching of thin plates (as assumed in classical plate theory). The assumptions are:
    - The normals to the mid-plane remain straight and normal to the midplane even after deformation.
    - The normals to the mid-plane do not change their lengths.
- The classical laminate theory is abbreviated as CLT. This theory is known as the classical laminated plate theory and abbreviated as CLPT.

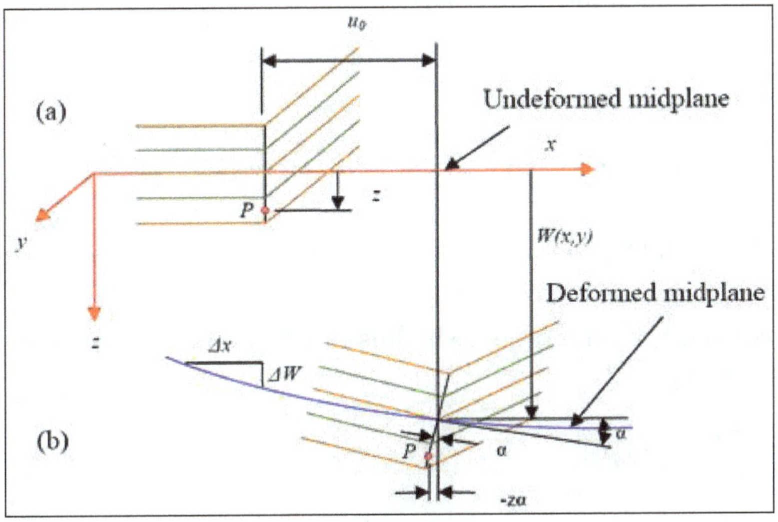

Plate Deformation

$$u(x,y,z) = u_0(x,y) - z\tan\alpha = u_0(x,y) - z\alpha = u_0(x,y) - z\frac{\partial W}{\partial x}.$$

Similarly, for the deformation in yz plane we can express the slope of the deformed midplane as.

Thus, the displacement of a generic point along y axis can be given as:

$$v(x,y,z) = v_o(x,y) - z\frac{\partial W}{\partial y},$$

Thus, the complete displacement field for a generic point in the laminate according to the classical laminate theory is given below:

$$\begin{aligned} u(x,y,z) &= u_o(x,y) - z\frac{\partial W}{\partial x} \\ v(x,y,z) &= v_o(x,y) - z\frac{\partial W}{\partial y} \\ w(x,y,z) &= w_o(x,y) \end{aligned} .$$

From the first assumption of the Kirchhoff-Love theory that the normals remain straight and normal to mid-plane even after deformation, results into zero transverse shear strains. Thus,

$$\gamma_{xz} = \gamma_{yz} = 0 .$$

Using the definitions of small strain, we can write the above equation as:

$$\begin{aligned} \gamma_{xz} &= \frac{\partial u}{\partial z} + \frac{\partial w}{\partial x} = 0 \\ \gamma_{yz} &= \frac{\partial v}{\partial z} + \frac{\partial w}{\partial y} = 0 . \end{aligned}$$

From the first of the above equation we can write:

$$\frac{\partial u}{\partial z} = -\frac{\partial w}{\partial x},$$

Integrating this with respect to z, we get:

$$u(x,y,z) = -z\frac{\partial w}{\partial z} + u_o(x,y),$$

Where $u_o(x,y)$ is a constant of integration which is function of x and y alone. Similarly, from the previous equation, we can get:

$$v(x,y,z) = -z\frac{\partial w}{\partial y} + v_o(x,y).$$

Strain Displacements Relations

The strain displacement relations for infinitesimal strains using the displacement field can be given as:

$$\epsilon_{xx} = \frac{\partial u}{\partial x} = \frac{\partial u_o}{\partial x} - z\frac{\partial^2 w}{\partial x^2},$$

$$\epsilon_{yy} = \frac{\partial v}{\partial y} = \frac{\partial v_o}{\partial y} - z\frac{\partial^2 w}{\partial y^2}$$

$$\gamma_{xy} = \frac{\partial u}{\partial y} + \frac{\partial v}{\partial x} = \frac{\partial u_o}{\partial y} + \frac{\partial v_o}{\partial x} - 2z\frac{\partial^2 w}{\partial x \partial y}.$$

The above equation can be written as:

$$\begin{Bmatrix} \epsilon_{xx} \\ \epsilon_{yy} \\ \gamma_{xy} \end{Bmatrix} = \begin{Bmatrix} \epsilon_{xx}^{(o)} \\ \epsilon_{yy}^{(o)} \\ \gamma_{xy}^{(o)} \end{Bmatrix} + z\begin{Bmatrix} k_{xx} \\ k_{yy} \\ k_{xy} \end{Bmatrix}$$

Or,

$$\left\{\epsilon^{(o)}\right\}_{xy} = \left\{\epsilon_{xx}^{(o)} \quad \epsilon_{yy}^{(o)} \quad \epsilon_{xy}^{(o)}\right\}^T = \left\{\frac{\partial u_o}{\partial x} \quad \frac{\partial v_o}{\partial y} \quad \frac{\partial u_o}{\partial y} + \frac{\partial v_o}{\partial x}\right\}^T$$

$$\{k\}_{xy} = \{k_{xx} \quad k_{yy} \quad k_{xy}\}^T = \left\{-\frac{\partial^2 w}{\partial x^2} \quad -\frac{\partial^2 w}{\partial y^2} \quad -2\frac{\partial^2 w}{\partial x \partial y}\right\}^T,$$

The terms k_{xx} and k_{yy} are the bending moment curvatures and k_{xy} is the twisting moment curvature.

Strain Displacements Relations

The strain displacement relations for infinitesimal strains using the displacement field can be given as:

$$\epsilon_{xx} = \frac{\partial u}{\partial x} = \frac{\partial u_o}{\partial x} - z\frac{\partial^2 w}{\partial x^2}$$

$$\epsilon_{yy} = \frac{\partial v}{\partial y} = \frac{\partial v_o}{\partial y} - z\frac{\partial^2 w}{\partial y^2}$$

$$\gamma_{xy} = \frac{\partial u}{\partial y} + \frac{\partial v}{\partial x} = \frac{\partial u_o}{\partial y} + \frac{\partial v_o}{\partial x} - 2z\frac{\partial^2 w}{\partial x \partial y}$$

The above equation can be written as:

$$\begin{Bmatrix} \epsilon_{xx} \\ \epsilon_{yy} \\ \gamma_{xy} \end{Bmatrix} = \begin{Bmatrix} \epsilon_{xx}^{(o)} \\ \epsilon_{yy}^{(o)} \\ \gamma_{xy}^{(o)} \end{Bmatrix} + z\begin{Bmatrix} k_{xx} \\ k_{yy} \\ k_{xy} \end{Bmatrix}$$

Or,

$$\{\in\}_{yy} = \{\varepsilon^{(o)}\}_{xy} + z\{k\}_{xy},$$

where $\{\in^{(o)}\}_{xy} = \{\in_{xx}^{(o)} \ \in_{yy}^{(o)} \ \gamma_{xy}^{(o)}\}^T = \left\{\frac{\partial u_o}{\partial x} \ \frac{\partial v_o}{\partial y} \ \frac{\partial u_o}{\partial y} + \frac{\partial v_o}{\partial x}\right\}^T$ are the midplane strains and $\{k\}_{xy} = \{k_{xx} \ k_{yy} \ k_{xy}\}^T = \left\{-\frac{\partial^2 w}{\partial x^2} \ -\frac{\partial^2 w}{\partial y^2} \ -2\frac{\partial^2 w}{\partial x \partial y}\right\}^T$ represents the midplane curvatures. The terms and are the bending moment curvatures and is the twisting moment curvature.

State of Stress in a Laminate

The stresses at any location can be calculated from the strains and lamina constitutive relations. It is assumed that the lamina properties are known. Hence, the constitutive equation for a kth lamina is known, that is, the reduced stiffness matrices (in principal material directions and global directions) are known. Thus, the stresses in k^{th} lamina can be given as:

$$\{\sigma\}_{xy}^k = [\bar{Q}]^k \{\in\}_{xy}^k.$$

Now, we can write the stresses as:

$$\{\sigma\}_{xy}^k = [\bar{Q}]^k \{\in^{(o)}\}_{xy} + [\bar{Q}]^k z\{k\}_{xy}.$$

In these equations, the strains are given at a z location where the stresses are required. It should be noted that the strains are continuous and vary linearly through the thickness. If we look at the stress distribution through the thickness it is clear that the stresses are not continuous through the thickness, because the stiffness is different for different laminae in thickness direction. In a lamina the stress varies linearly. The slope of this variation in a lamina depends upon its moduli. However, at the interface of two adjacent laminae there is a discontinuity in the stresses. The same thing is depicted in below figure with three layers.

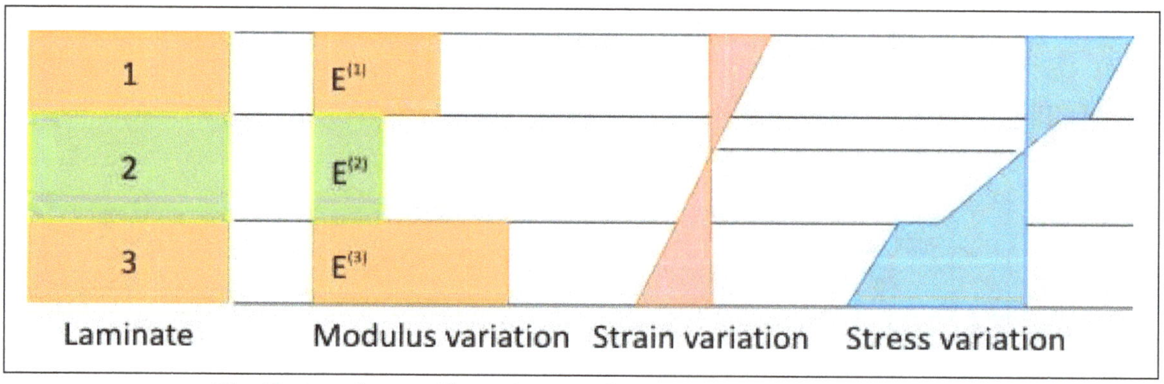

Elucidation of stress discontinuity at lamina interfaces in a laminate.

Inplane Resultant Forces

The inplane forces per unit length are defined as:

$$N_{xx} = \int_{-H}^{H} \sigma_{xx} dz, \quad N_{yy} = \int_{-H}^{H} \sigma_{yy} dz, \quad N_{xy} = \int_{-H}^{H} \tau_{xy} dz,$$

Or these can be written as:

$$N_{xy} = \int_{-H}^{H} \sigma_{xx} dz,$$

$$\{N_{xy}\} = \sum_{k=1}^{N_{Lay}} \int_{-z_{k-1}}^{zk} [\bar{Q}]^k \{\in^{(o)}\}_{xy} dz + \sum_{k=1}^{N_{Lay}} \int_{-z_{k-1}}^{zk} [\bar{Q}]^k \{k\}_{xy} z\, dz.$$

Now recall that the midplane strains $\{\in^{(o)}\}_{xy}$ and the curvatures $\{\ \}_{xy}$ are independent of z location. The reduced transformed stiffness matrix $[\bar{Q}]$ is function of thickness and constant over a given lamina thickness. Now we can replace the integration over the laminate thickness as sum of the integrations over individual lamina thicknesses. Thus, Equation can be written as:

$$\{N_{xy}\} = \sum_{k=1}^{N_{Lay}} \int_{-z_{k-1}}^{zk} [\bar{Q}]^k \{\in^{(o)}\}_{xy} dz + \sum_{k=1}^{N_{Lay}} \int_{-z_{k-1}}^{zk} [\bar{Q}]^k \{k\}_{xy} z\, dz,$$

Here, is the total number of layers in the laminate. This equation can be written as:

$$[A] = \sum_{k=1}^{N_{Lay}} [\bar{Q}]^k (z_k - z_{k-1}) \text{ and } [B] = \frac{1}{2} \sum_{k=1}^{N_{Lay}} [\bar{Q}]^k (z_k^2 - z_{k-1}^2).$$

The matrix [A] represents the in-plane stiffness, that is, it relates the in-plane forces with mid-plane strains and the matrix [B] represents the bending stiffness coupling, that is, it relates the in-plane forces with mid-plane curvatures.

It should be noted that the matrices [A] and [B] are symmetric as the matrix $[\bar{Q}]$ is also symmetric for each lamina in the laminate. The resultant in-plane forces are shown.

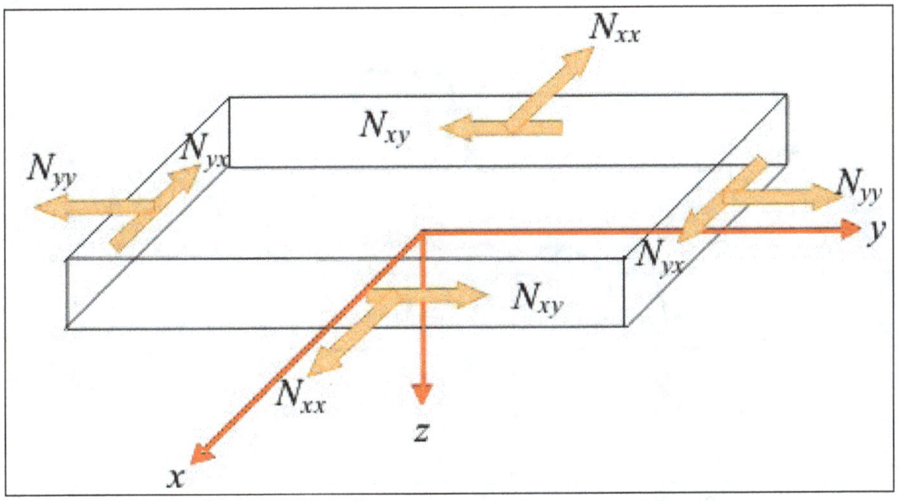

In plane resultant forces per unit length on a laminate.

Resultant Moments

The resultant moments per unit length are defined as:

$$M_{xx} = \int_{-H}^{H} \sigma_{xx} z\, dz, \quad M_{yy} = \int_{-H}^{H} \sigma_{yy} z\, dz, \quad M_{xy} = \int_{-H}^{H} \tau_{xy} z\, dz,$$

Or these can be written as:

$$\{M\}_{xy} = \int_{-H}^{H} \{\sigma\}_{xy} z\, dz,$$

Now, we can write:

$$\{M\}_{xy} = \int_{-H}^{H} \{\bar{Q}\}^{k} \{\in^{(o)}\}_{xy} z\, dz + \int_{-H}^{H} \{\bar{Q}\}^{k} \{k\}_{xy} z^2\, dz,$$

Now, with the same justification as given, we can write the above equation as:

$$\{M_{xy}\} = \sum_{k=1}^{N_{Lay}} \int_{-z_{k-1}}^{zk} \left[\bar{Q}\right]^{k} \{\in^{(o)}\}_{xy} z\, dz + \sum_{k=1}^{N_{Lay}} \int_{-z_{k-1}}^{zk} \left[\bar{Q}\right]^{k} \{k\}_{xy} z^2\, dz.$$

This can be written as:

$$\{M\}_{xy} = [B]\{\in^{(o)}\}_{xy} + [D]\{k\}_{xy}.$$

Where,

$$[D] = \frac{1}{3} \sum_{k=1}^{N_{Lay}} \left[\bar{Q}\right]^{k} \left(z_k^3 - z_{k-1}^3\right).$$

The matrix [D] represents the bending stiffness, that is, it relates resultant moments with mid-plane curvatures. Again, the matrix [D] is also symmetric. Further, it is important to note that the matrix [B] relates the resultant moments with mid-plane curvatures as well.

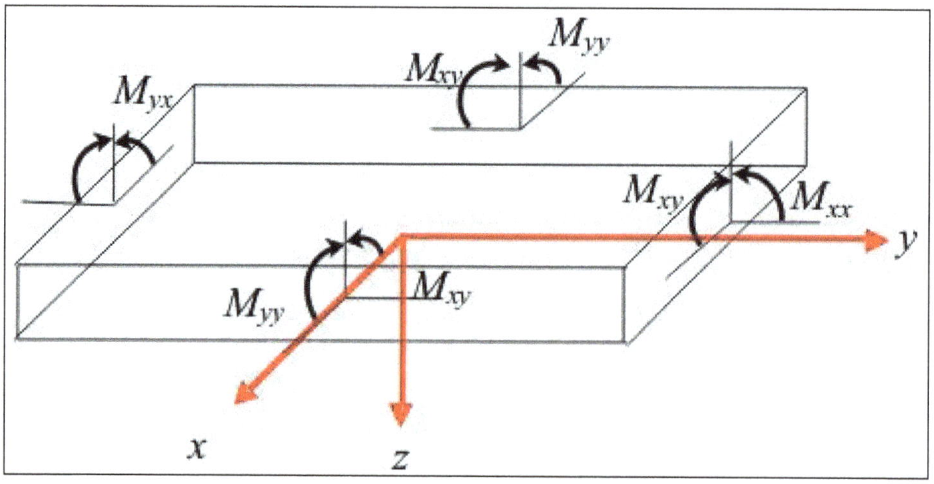

Resultant moments per unit length on a laminate.

Properties of Composites

Isostrain Rules of Mixtures

If both the fibres and matrix are aligned parallel to the loading direction, the deformation of both phases will be the same (assuming there is no delamination at the fibre-matrix interface). This isostrain condition provides the upper bound for composite strength, and is determined by the rule of mixtures:

$$E_C = \sum_{i=1} V_i E_i.$$

Where E_C is the effective composite Young's modulus, and V_i and E_i are the volume fraction and Young's moduli, respectively, of the composite phases.

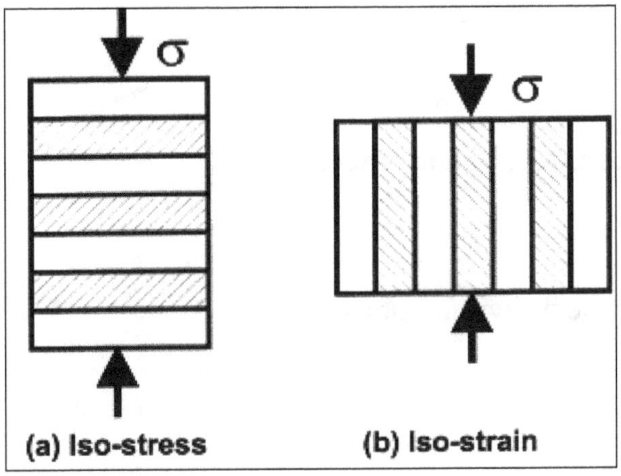

Figure a) shows the isostress condition where the composite materials are perpendicular to the applied force and b) is the isostrain condition that has the layers parallel to the force.

For example, a composite material made up of α and β phases as shown in the figure to the right under isostrain, the Young's modulus would be as follows:

$$E_C = V_\alpha E_\alpha + V_\beta E_\beta.$$

Where V_α and V_β are the respective volume fractions of each phase. This can be derived by considering that in the isostrain case,

$$\epsilon_C = \epsilon_\alpha = \epsilon_\beta = \epsilon.$$

Assuming that the composite has a uniform cross section, the stress on the composite is a weighted average between the two phases,

$$\sigma_C = \sigma_\alpha V_\alpha + \sigma_\beta V_\beta.$$

The stresses in the individual phases are given by Hooke's Law,

$$\sigma_\beta = E_\beta \epsilon$$
$$\sigma_\alpha = E_\alpha \epsilon$$

Combining these equations gives that the overall stress in the composite is:

$$\sigma_C = E_\alpha V_\alpha \epsilon + E_\beta V_\beta \epsilon = (E_\alpha V_\alpha + E_\beta V_\beta)\epsilon .$$

Then it can be shown that:

$$E_C = (E_\alpha V_\alpha + E_\beta V_\beta).$$

Isostress Rule of Mixtures

The lower bound is dictated by the isostress condition, in which the fibres and matrix are oriented perpendicularly to the loading direction:

$$\sigma_C = \sigma_\alpha = \sigma_\beta = \sigma ,$$

and now the strains become a weighted average:

$$\epsilon_C = \epsilon_\alpha V_\alpha + \epsilon_\beta V_\beta .$$

Rewriting Hooke's Law for the individual phases:

$$\epsilon_\beta = \frac{\sigma}{E_\beta},$$

$$\epsilon_\alpha = \frac{\sigma}{E_\alpha}.$$

This leads to:

$$\epsilon_c = V_\beta \frac{\sigma}{\epsilon_\beta} + V_\alpha \frac{\sigma}{\epsilon_\alpha} = (\frac{V_\alpha}{\epsilon_\alpha} + \frac{V_\beta}{\epsilon_\beta})\sigma .$$

From the definition of Hooke's Law:

$$\frac{1}{E_C} = \frac{V_\alpha}{E_\alpha} + \frac{V_\beta}{E_\beta},$$

and in general,

$$\frac{1}{E_C} = \sum_{i=1} \frac{V_i}{E_i}.$$

Following the example above, if one had a composite material made up of α and β phases under isostress conditions as shown in the figure to the right, the composition Young's modulus would be:

$$E_C = (E_\alpha E_\beta)/(V_\alpha E_\beta + V_\beta E_\alpha).$$

The isostrain condition implies that under an applied load, both phases experience the same strain but will feel different stress. Comparatively, under isostress conditions both phases will feel the same stress but the strains will differ between each phase. A generalized equation for any loading condition between isostrain and isostress can be written as:

$$(X_c)^n = V_m(X_m)^n + V_r(X_r)^n,$$

Where X is a material property such as modulus or stress, c, m, and r stand for the properties of the composite, matrix, and reinforcement materials respectively, and n is a value between 1 and −1.

The above equation can be further generalized beyond a two phase composite to an m-component system:

$$(X_c)^n = \sum_{i=1}^{m} V_i(X_i)^n.$$

Though composite stiffness is maximized when fibres are aligned with the loading direction, so is the possibility of fibre tensile fracture, assuming the tensile strength exceeds that of the matrix. When a fibre has some angle of misorientation θ, several fracture modes are possible. For small values of θ the stress required to initiate fracture is increased by a factor of $(\cos \theta)^{-2}$ due to the increased cross-sectional area (A cos θ) of the fibre and reduced force (F/cos θ) experienced by the fibre, leading to a composite tensile strength of $\sigma_{parallel}/\cos^2\theta$ where $\sigma_{parallel}$ is the tensile strength of the composite with fibres aligned parallel with the applied force.

Intermediate angles of misorientation θ lead to matrix shear failure. Again the cross sectional area is modified but since shear stress is now the driving force for failure the area of the matrix parallel to the fibres is of interest, increasing by a factor of 1/sin θ. Similarly, the force parallel to this area again decreases (F/cos θ) leading to a total tensile strength of $T_{my}/\sin\theta\cos\theta$ where T_{my} is the matrix shear strength.

Finally, for large values of θ (near π/2) transverse matrix failure is the most likely to occur, since the fibres no longer carry the majority of the load. Still, the tensile strength will be greater than for the purely perpendicular orientation, since the force perpendicular to the fibres will decrease by a factor of 1/sin θ and the area decreases by a factor of 1/sin θ producing a composite tensile strength of $\sigma_{perp}/\sin^2\theta$ where σ_{perp} is the tensile strength of the composite with fibres align perpendicular to the applied force.

The majority of commercial composites are formed with random dispersion and orientation of the strengthening fibres, in which case the composite Young's modulus will fall between the isostrain and isostress bounds. However, in applications where the strength-to-weight ratio is engineered to be as high as possible (such as in the aerospace industry), fibre alignment may be tightly controlled. Panel stiffness is also dependent on the design of the panel. For instance, the fibre reinforcement and matrix used, the method of panel build, thermoset versus thermoplastic, and type of weave.

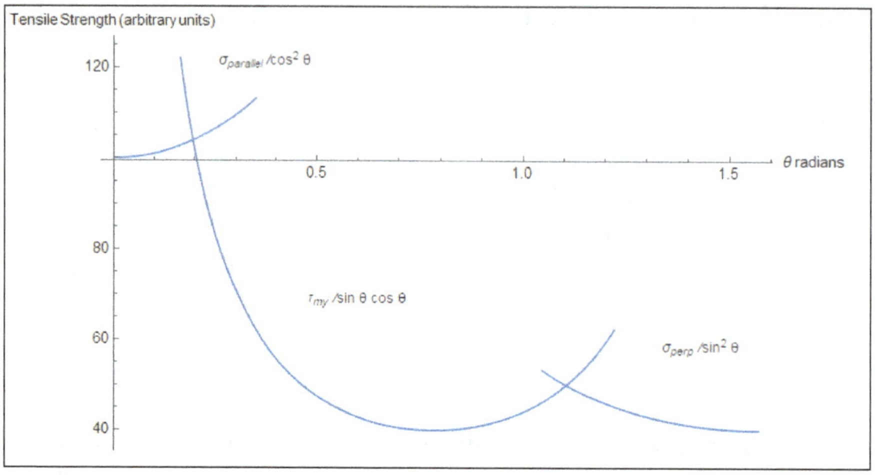

The graph depicts the three fracture modes a composite material may experience depending on the angle of misorientation relative to aligning fibres parallel to the applied stress.

In contrast to composites, isotropic materials (for example, aluminium or steel), in standard wrought forms, possess the same stiffness typically despite the directional orientation of the applied forces and/or moments. The relationship between forces/moments and strains/curvatures for an isotropic material can be described with the following material properties: Young's Modulus, the Shear Modulus and the Poisson's ratio, in relatively simple mathematical relationships. For the anisotropic material, it needs the mathematics of a second-order tensor and up to 21 material property constants.

For the special case of orthogonal isotropy, there are three distinct material property constants for each of Young's Modulus, Shear Modulus and Poisson's ratio—a total of 9 constants to express the relationship between forces/moments and strains/curvatures. Techniques that take benefit of the materials' anisotropic properties involve mortise and tenon joints.

Coefficient of Thermal Expansion

1. Coefficient of Thermal Expansion (CTE) in longitudinal direction (along the fibers):

$$\alpha_{cl} = (\alpha_m * E_m * V_m + \alpha_f * E_f * V_f) / (E_m * V_m + E_f * V_f)$$

Where:

- α_{cl}, α_m, α_f – CTE of composite in longitudinal direction, matrix and dispersed phase (fiber) respectively.
- E_m,E_f – Modulus of elasticity of matrix and dispersed phase (fiber) respectively.

2. Coefficient of Thermal Expansion (CTE) in transverse direction (perpendicular to the fibers):

$$\alpha_{ct} = (1 + P_m)\alpha_m * V_m + \alpha_f * V_f$$

Where:

- P_m – Poisson ratio of matrix.

Poisson's ratio is the ratio of transverse contraction strain to longitudinal extension strain in the direction of applied force.

Comparison between Metals and Composites

The physical characteristics of composites and metals are significantly different. Because composites are highly anisotropic, their in-plane strength and stiffness are usually high and directionally variable, depending on the orientation of the reinforcing fibers. Properties that do not benefit from this reinforcement (at least for polymer matrix composites) are comparatively low in strength and stiffness—for example, the through-the thickness tensile strength where the relatively weak matrix is loaded rather than the high-strength fibers. Figure below shows the low through-the-thickness strength of a typical composite laminate compared with aluminum.

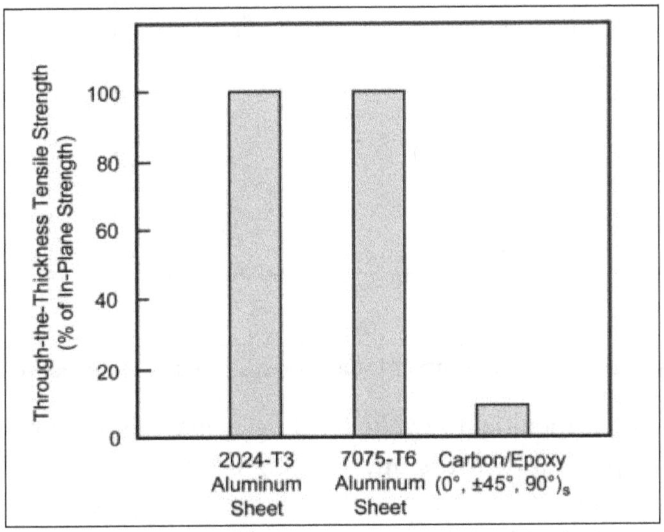

Comparison of through-the-thickness tensile strength of a composite laminate with aluminum alloy sheet.

Table: Composites versus metals comparison.

Condition	Comparative behavior relative to metals
Load-strain relationship	More linear strain to failure
Notch sensitivity	
Static	Greater sensitivity
Fatigue	Less sensitivity
Transverse properties	Weaker
Mechanical property variability	Higher
Fatigue strength	Higher
Sensitivity to hydrothermal environment	Greater
Sensitivity to corrosion	Much less
Damage growth mechanism	In-plane delamination instead of through thickness cracks

Metals typically have reasonable ductility, continuing to elongate or compress considerably when they reach a certain load (through yielding) without picking up more load and without failure. Two important benefits of this ductile yielding are that (1) it provides for local load relief by distributing excess load to an adjacent material or structure; therefore, ductile metals have a great capacity to provide relief from stress concentrations when statically loaded; and (2) it provides great energy-absorbing capability (indicated by the area under a stress-strain curve). As a result, when impacted, a metal structure typically deforms but does not actually fracture. In contrast, composites are relatively brittle. Figure below shows a comparison of typical tensile stress-strain curves for two materials. The brittleness of the composite is reflected in its poor ability to tolerate stress concentrations. The characteristically brittle composite material has poor ability to resist impact damage without extensive internal matrix fracturing.

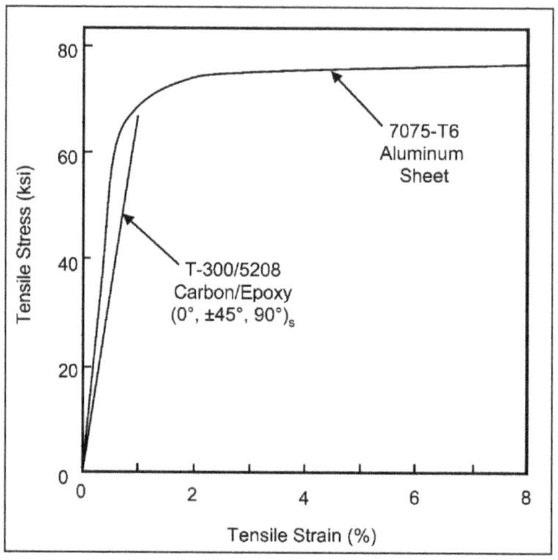

Comparison of typical stress-strain curves for a composite laminate and aluminum alloy sheet.

The response of damaged composites to cyclic loading is also significantly different from that of metals. The ability of composites to withstand cyclic loading is far superior to that of metals, in contrast to the poor composite static strength when it has damage or defects. Figure below shows a comparison of the normalized notched specimen fatigue response of a common 7075- T6 aluminum aircraft metal and a carbon/epoxy laminate. The fatigue strength of the composite is much higher relative to its static or residual strength. The static or residual strength requirement for structures is typically much higher than the fatigue requirement. Therefore, because the fatigue threshold of composites is a high percentage of their static or damaged residual strength, they are usually not fatigue critical. In metal structures, fatigue is typically a critical design consideration.

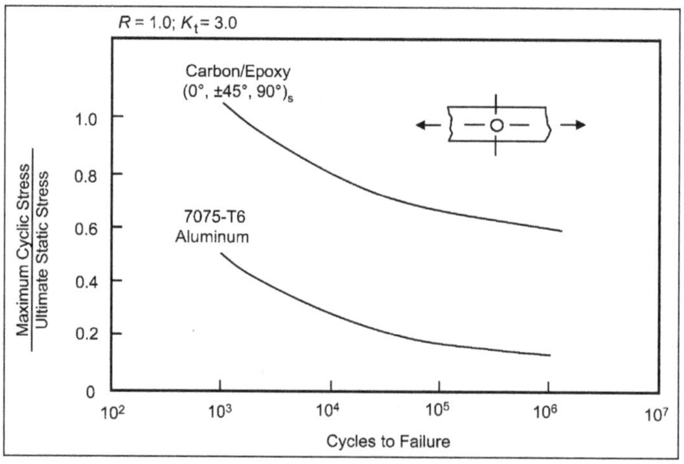

Comparative notched fatigue strength of composite laminate and aluminum alloy sheet.

References

- Introduction-of-Composites-105108124: ac.in, Retrieved 02, July 2020
- Composite-Interphases: pslc.ws, Retrieved 22, April 2020
- Composite-materials: mrcet.com, Retrieved 15, January 2020

2

Types of Matrix Composite

A few of the different types of matrix composites are metal matrix composite, ceramic matrix com-posite, polymer matrix composite and carbon matrix composite. A metal matrix composite has at least one constituent part which is a metal. The topics elaborated in this chapter will help in gain-ing a better perspective about these types of matrix composites.

Metal Matrix Composite

Metal Matrix Composites are composed of a metallic matrix (Al, Mg, Fe, Cu etc.) and a dispersed ceramic (oxide, carbides) or metallic phase(Pb, Mo, W etc.). Ceramic reinforcement may be silicon carbide, boron, alumina, silicon nitride, boron carbide, boron nitride etc. whereas Metallic Rein-forcement may be tungsten, beryllium etc. MMCs are used for Space Shuttle, commercial airliners, electronic substrates, bicycles, automobiles, golf clubs and a variety of other applications. From a material point of view, when compared to polymer matrix composites, the advantages of MMCs lie in their retention of strength and stiffness at elevated temperature, good abrasion and creep resistance properties. Most MMCs are still in the development stage or the early stages of produc-tion and are not so widely established as polymer matrix composites. The biggest disadvantages of MMCs are their high costs of fabrication, which has placed limitations on their actual applica-tions. There are also advantages in some of the physical attributes of MMCs such as no significant moisture absorption properties, non-inflammability, low electrical and thermal conductivities and resistance to most radiations. MMCs have existed for the past 30 years and a wide range of MMCs have been studied.

Compared to monolithic metals, MMCs have:

- Higher strength-to-density ratios.
- Higher stiffness-to-density ratios.
- Better fatigue resistance.
- Better elevated temperature properties:
    - Higher strength.
    - Lower creep rate.
- Lower coefficients of thermal expansion.
- Better wear resistance.

The advantages of MMCs over polymer matrix composites are:

- Higher temperature capability.
- Fire resistance.
- Higher transverse stiffness and strength.
- No moisture absorption.
- Higher electrical and thermal conductivities.
- Better radiation resistance.
- No out gassing.
- Fabric ability of whisker and particulate-reinforced MMCs with conventional metalworking equipment.

Some of the disadvantages of MMCs compared to monolithic metals and polymer matrix composites are:

- Higher cost of some material systems.
- Relatively immature technology.
- Complex fabrication methods for fiber-reinforced systems (except for casting).
- Limited service experience.

Numerous combinations of matrices and reinforcements have been tried since work on MMC began in the late 1950s. However, MMC technology is still in the early stages of development, and other important systems undoubtedly will emerge. Numerous metals have been used as matrices. The most important have been aluminum, titanium, magnesium, and copper alloys and superalloys. The most important MMC systems are:

- Aluminum matrix:
    - Continuous fibers: boron, silicon carbide, alumina, graphite.
    - Discontinuous fibers: alumina, alumina-silica.
    - Whiskers: silicon carbide.
    - Particulates: silicon carbide, boron carbide.
- Magnesium matrix:
    - Continuous fibers: graphite, alumina.
    - Whiskers: silicon carbide.
    - Particulates: silicon carbide, boron carbide.

- Titanium matrix:
    - Continuous fibers: silicon carbide, coated boron.
    - Particulates: titanium carbide.
- Copper matrix:
    - Continuous fibers: graphite, silicon carbide.
    - Wires: niobium-titanium, niobium-tin.
    - Particulates: silicon carbide, boron carbide, titanium carbide.

Ceramic Matrix Composite

Ceramic matrix composites (CMCs) have a ceramic matrix such as alumina calcium alumino silicate reinforced by fibers such as carbon or silicon carbide. Advantages of CMCs include high strength, hardness, high service temperature limits for ceramics, chemical inertness, and low density. However, ceramics by themselves have low fracture toughness. Under tensile or impact loading, they fail catastrophically. Reinforcing ceramics with fibers, such as silicon carbide or carbon, increases their fracture toughness because it causes gradual failure of the composite. This combination of a fiber and ceramic matrix makes CMCs more attractive for applications in which high mechanical properties and extreme service temperatures are desired. Typical mechanical properties of ceramic matrix composites are given in table below.

One of the most common methods to manufacture ceramic matrix composites is called the hot pressing method. Glass fibers in continuous tow are passed through slurry consisting of powdered matrix material, solvent such as alcohol, and an organic binder. The tow is then wound on a drum and dried to form prepreg tapes. The prepreg tapes can now be stacked to make a required laminate. Heating at about 932 °F (500 °C) burns out the binder. Hot pressing at high temperatures in excess of 1832 °F (1000 °C) and pressures of 1 to 2 ksi (7 to 14 MPa) follows this.

Property	Units	SiC/LAS	Sic/CAS	Steel	Aluminum
System of units: USCS					
Specific gravity	-	2.1	2.5	7.8	2.6
Young's Modulus	Msi	13	17.55	30.0	10.0
Ultimate tensile strength	ksi	72	58.0	94.0	34.0
Coefficient of thermal expansion	μin./in./°F	2	2.5	6.5	12.8
System of units: SI					
Specific Gravity	-	2.1	2.5	7.8	2.6
Young's Modulus	GPa	89.63	121	206.8	68.95
Ultimate tensile strength	MPa	496.4	400	648.1	234..4
Coefficient of thermal expansion	μm/m/°C	3.6	4.5	11.7	23

Ceramic matrix composites are finding increased application in high-temperature areas in which metal and polymer matrix composites cannot be used. This is not to say that CMCs are not attractive otherwise, especially considering their high strength and modulus, and low density. Typical applications include cutting tool inserts in oxidizing and high-temperature environments.

Polymer Matrix Composite

Polymers make ideal materials as they can be processed easily, possess lightweight, and desirable mechanical properties. It follows, therefore, that high temperature resins are extensively used in aeronautical applications.

Two main kinds of polymers are thermosets and thermoplastics. Thermosets have qualities such as a well-bonded three-dimensional molecular structure after curing. They decompose instead of melting on hardening. Merely changing the basic composition of the resin is enough to alter the conditions suitably for curing and determine its other characteristics. They can be retained in a partially cured condition too over prolonged periods of time, rendering Thermosets very flexible. Thus, they are most suited as matrix bases for advanced conditions fiber reinforced composites. Thermosets find wide ranging applications in the chopped fiber composites form particularly when a premixed or moulding compound with fibers of specific quality and aspect ratio happens to be starting material as in epoxy, polymer and phenolic polyamide resins.

Thermoplastics have one- or two-dimensional molecular structure and they tend to at an elevated temperature and show exaggerated melting point. Another advantage is that the process of softening at elevated temperatures can reversed to regain its properties during cooling, facilitating applications of conventional compress techniques to mould the compounds.

Resins reinforced with thermoplastics now comprised an emerging group of composites. The theme of most experiments in this area to improve the base properties of the resins and extract the greatest functional advantages from them in new avenues, including attempts to replace metals in die-casting processes. In crystalline thermoplastics, the reinforcement affects the morphology to a considerable extent, prompting the reinforcement to empower nucleation. Whenever crystalline or amorphous, these resins possess the facility to alter their creep over an extensive range of temperature. But this range includes the point at which the usage of resins is constrained, and the reinforcement in such systems can increase the failure load as well as creep resistance. Figure below shows kinds of thermoplastics.

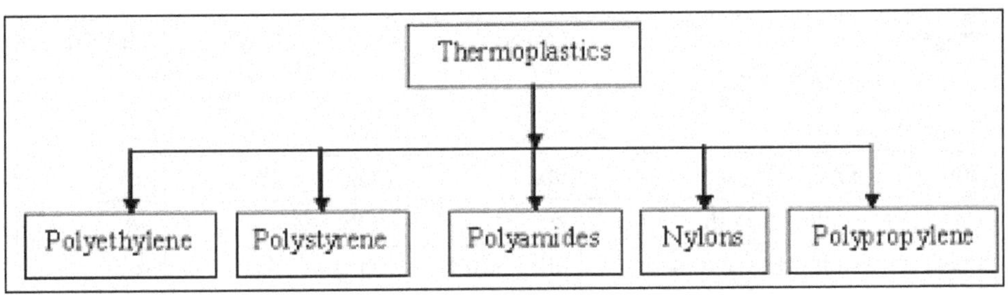

Thermoplastics

A small quantum of shrinkage and the tendency of the shape to retain its original form are also to be accounted for. But reinforcements can change this condition too. The advantage of thermoplastics systems over thermosets are that there are no chemical reactions involved, which often result in the release of gases or heat. Manufacturing is limited by the time required for heating, shaping and cooling the structures.

Thermoplastics resins are sold as moulding compounds. Fiber reinforcement is apt for these resins. Since the fibers are randomly dispersed, the reinforcement will be almost isotropic. However, when subjected to moulding processes, they can be aligned directionally.

Carbon Matrix Composite

Carbon–carbon composites use carbon fibers in a carbon matrix. These composites are used in very high-temperature environments of up to 6000 °F (3315 °C), and are 20 times stronger and 30% lighter than graphite fibers.

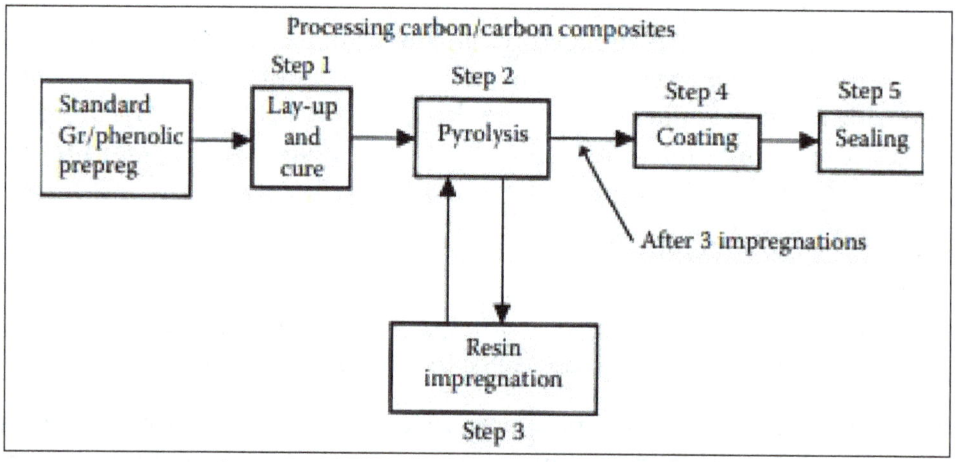

Schematic of processing carbon–carbon composites.

Carbon is brittle and flaw sensitive like ceramics. Reinforcement of a carbon matrix allows the composite to fail gradually and also gives advantages such as ability to withstand high temperatures, low creep at high temperatures, low density, good tensile and compressive strengths, high fatigue resistance, high thermal conductivity, and high coefficient of friction. Drawbacks include high cost, low shear strength, and susceptibility to oxidations at high temperatures. Typical properties of carbon–carbon composites are given in table below.

Property	Units	C-C	Steel	Aluminum
System of units: USCS				
Specific gravity	-	1.68	7.8	2.6
Young's Modulus	Msi	1.95	30	10
Ultimate tensile strength	ksi	5.180	94	40
Coefficient of thermal expansion	µin./in./°F	1.11	6.5	12.8
System of units: SI				

Specific Gravity	-	1.68	7.8	2.6
Young's Modulus	GPa	13.5	206.8	68.95
Ultimate strength	MPa	35.7	648.1	234..4
Coefficient of thermal expansion	μm/m/°C	2.0	11.7	23

A typical method for manufacturing carbon–carbon composites is called low-pressure carbonization. A graphite cloth is taken, impregnated by resin (such as phenolic, pitch, and furfuryl ester), and laid up in layers. It is laid in a mold, cured, and trimmed. The part is then pyrolized, converting the phenolic resin to graphite. The composite is then impregnated by furfuryl alcohol. The process drives off the resin and any volatiles. The process is repeated three or four times until the level of porosity is reduced to an acceptable level. Each time, this process increases its modulus and strength. Because carbon–carbon composites oxidize at temperatures as low as 842 °F (450 °C), an outer layer of silicon carbide may be deposited.

References

- Metal-Matrix-Composite-10404002: nitrkl.ac.in, Retrieved 21, August 2020
- Mechanics-of-composite-materials: iut.ac.ir, Retrieved 03, February 2020
- Introduction-of-Composites-105108124: ac.in, Retrieved 11, June 2020

3

Fabrication Techniques

There are various techniques which can be used for the fabrication of composite materials such as open moulding, closed moulding, autoclave moulding, lanxide process, Z-pinning, stitching technology and tufting. This chapter closely examines these fabrication techniques to provide an extensive understanding of the subject.

Manufacturing of composite involves manufacturing of fiber preforms and then reinforcing these fibers with the matrix material by various techniques. Fiber preforms involve weaving, knitting, braiding, and stitching of fibers in long sheets or mat structure. Preforms are used to achieve a high level of automation with the assistance of robotics, which offers control over the fiber angle and the fiber content on every zone of the part to be molded.

Open Moulding

In open moulding, raw materials (resins and fiber reinforcements) are exposed to air as they cure or harden. Open moulding utilizes different processes, including hand lay-up, spray-up, casting, and filament winding. Prepregs are a combination of fibers and uncured resin, which are pre-impregnated with thermoplastic or a thermoset resin material that only needs the temperature to be activated. These prepregs are ready-to-use materials where the readily impregnated layers are cut and laid down into the open mold.

Hand Lay-Up

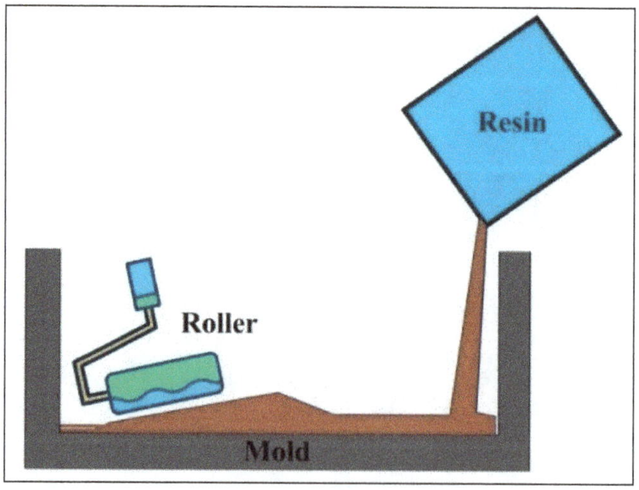

Hand layup process.

Figure below shows the hand lay-up, which is the most common and widely used open mold composite manufacturing process. Initially, fiber preforms are placed in a mold where a thin layer of anti-adhesive coat is applied for easy extraction. The resin material is poured or applied using a brush on a reinforcement material. The roller is used to force the resin into the fabrics to ensure an enhanced interaction between the successive layers of the reinforcement and the matrix materials.

Spray Up

In the spray-up process, the operator controls thickness and consistency, therefore the process is more operator dependent than hand lay-up. Although production volume per mold is low, it is feasible to produce substantial production quantities using multiple molds. This process uses simple, low cost tooling and simple processing. Portable equipment permits on-site fabrication with virtually no part size limitations. The process may be automated.

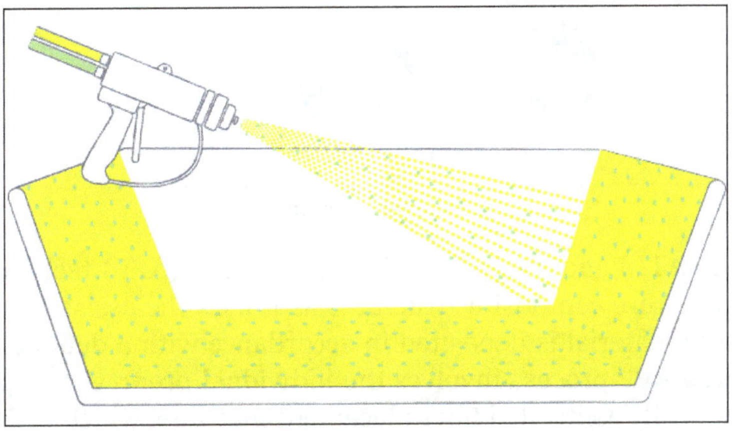

As with hand lay-up, gel coat is first applied to the mold and allowed to cure. Continuous strand glass roving and initiated resin are then fed through a chopper gun, which deposits the resin-saturated "chop" on the mold. The laminate is then rolled to thoroughly saturate the glass strands and compact the chop. Additional layers of chop laminate are added as required for thickness. Roll stock reinforcements, such as woven roving or knitted fabrics, can be used in conjunction with the chopped laminates. Core materials of the same variety as used in hand lay-up are easily incorporated.

Molds

These are the same molds as hand lay-up: simple, single cavity molds of fiberglass composites construction. Molds can range from small to very large and are low cost in the spectrum of composites molds.

Filament Welding

It is a continuous process that offers self-automation, which leads to reduced cost. Filament winding is useful to create axisymmetric, as well as some non-axisymmetric, composite parts, such as pipe bends. Driven by several pulleys, continuous prepreg sheets, rovings, and monofilament are made to pass through a resin bath and collected over a rotating mandrel, as displayed in Figure below. Then, after applying sufficient layers, mandrel, which has the desired shape of the product, is set for curing at the room temperature. Recently developed robotic filament winding (RFW) technique is provided with an industrial robot equipped with a feed and deposition system. It yields advantages over process control, repeatability, and manufacturing time by replacing a human operator.

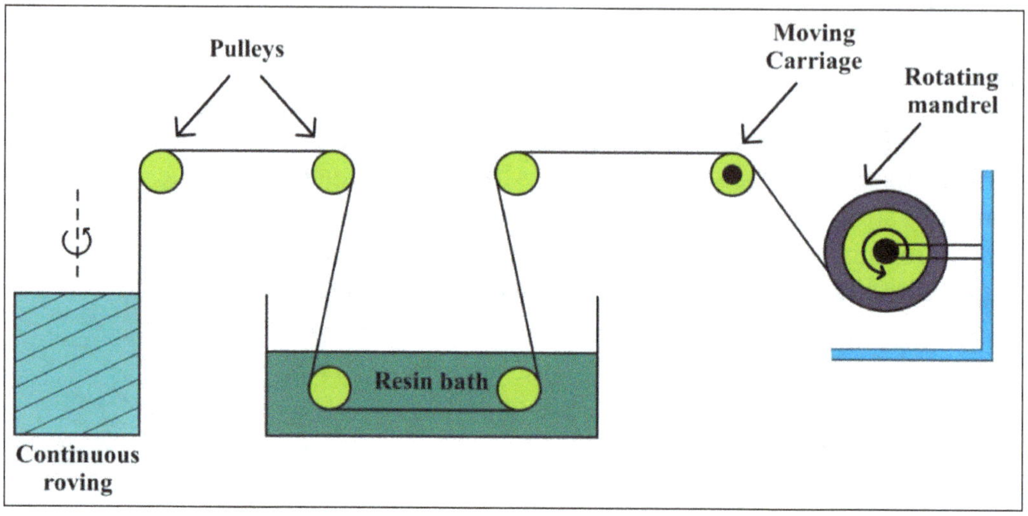

Filament winding

Automated tape layup (ATL) and automated fiber placement (AFP) techniques are efficient for large, flat, or single curvature composite structures as it uses the assistance of a multiaxis articulating robot, where the material is deposited in accordance with a defined computer numerical control (CNC) path. The AFP process involves the individual prepreg lay-up of laminates onto a mandrel using a numerically controlled fiber placement machine, which is then further pulled off by holding spools. Composite structures are fabricated quickly and accurately, but the expenses in employing required specialized equipment keep these technologies out of reach for small to medium scale manufacturers.

Closed Moulding

In closed-moulding, raw materials (fibers and resin) cure inside a two-sided mold or within a vacuum bag (shut off from air). Closed-moulding processes are usually automated and require special equipment, so they're mainly used in large plants that produce huge volumes of material—up to 500,000 parts a year.

Vacuum Bag Moulding

Vacuum bag moulding uses a flexible film made of a material such as nylon polyethylene or polyvinyl alcohol (PVA) to enclose and seal the part from the outside air. Many times, the vacuum bag moulding technique is performed with the assistance of the hand lay-up technique. Laminate is first made by using the hand lay-up technique, and then after it is placed between the vacuum bag and the mold to ensure fair infusion of fibers into the matrix material. The air between the mold and the vacuum bag is then drawn out by a vacuum pump while atmospheric pressure compresses the part. The process can be well understood by Figure below. Hierarchical composites were prepared with multiscale reinforcements of carbon fibers using a vacuum bagging process, which eliminated chances of detectable porosity and improper impregnation of dual reinforcements, with increases in flexural and interlaminar shear properties by 15% and 18%, respectively.

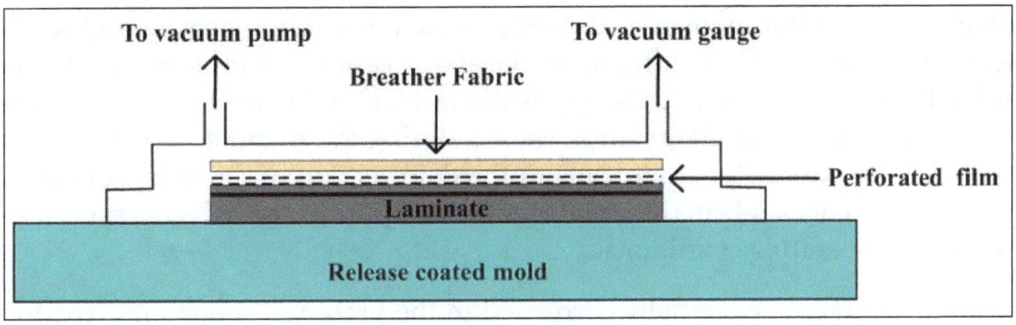

Vacuum bag moulding process.

Pressure Bag Moulding

This process is related to vacuum bag moulding in exactly the same way as it sounds. A solid female mould is used along with a flexible male mould. The reinforcement is place inside the female mould with just enough resin to allow the fabric to stick in place. A measured amount of resin is then liberally brushed indiscriminately into the mould and the mould is then clamped to a machine that contains the male flexible mould. The flexible male membrane is then inflated with heated compressed air or possibly steam. The female mould can also be heated. Excess resin is forced out along with trapped air. This process is extensively used in the production of composite helmets due to the lower cost of unskilled labor. Cycle times for a helmet bag moulding machine vary from 20 to 45 minutes, but the finished shells require no further curing if the moulds are heated.

Pressure bag moulding is similar to the vacuum bag moulding method except that air pressure, usually 30 to 50 psi, is applied to a rubber bag, or sheet that covers the laid up composite to force out entrapped air and excess resin. Pressurized steam may be used instead, to accelerate the cure. Cores and inserts can be used with the process, and undercuts are practical, but only female and split molds can be used to make items such as tanks, containers, and wind turbine blades.

Resin Transfer Moulding

The Resin Transfer Moulding (RTM) is one of the most promising technology available today. RTM is capable of making large complex three-dimensional part with high mechanical performance, tight dimensional tolerance and high surface finish. A good design by RTM leads to fabricate three-dimensional near-net-shape complex parts, offering production of cost-effective structural parts in medium-volume quantities using low cost tooling. In addition to these advantages, the problems of the joints, typical of the metal structures, can be eliminated by integration of inserts.

The final performances of a composite depend not only on the choice of the matrix and the fiber but also on the manufacturing process by which they are made. Since the starting of the composite life, the presence of imperfections due to manufacturing must be considered. Such imperfections can be already damage for the manufactured composite piece or lead to the damage quickly. The damage for composites can be defined as a change in the microstructure of the material that causes deterioration in the structural behavior of the component and sometimes its collapse. The damage in a composite structure can occur at the level of fibers and matrix as well.

The most common damage to the fibers is the interruption of their continuity. For example, fibers that are subject to load tend to align again inducing states of compression and tension on the matrix. These states may cause, in addition to a local decrease of the properties of the lamina, the breaking of the fibers themselves and the gap between fiber and matrix. In the assumption of an optimal stratification, fiber misalignment is caused during the manufacturing process by a bad balance of the process parameters that leads to the deformation of the fiber bundles. The fibers may also be distributed unevenly in the volume of the composite, and this generates intra-laminar shear stresses under operating conditions.

In the matrix, the damage is essentially correlated to the presence of porosity. The formation of microvoids between fibers and dry spots are potential starting points for propagating cracks or delaminations. The response of the component to delamination depends not only on the compatibility and surface tension between fibers and matrix but mainly on the compaction, and therefore also on the impregnation phases during the manufacturing process.

Comparing RTM with traditional manufacturing process applied in the aerospace industries (i.e. autoclave), the RTM technique results in a suitable alternative to the prepreg approach permitting high finish quality and controlled fiber directions. RTM reduces voids compared to hand lay-up so increasing component mechanical properties. In addition, component design by RTM process can compete with metal one when prepreg cannot be applied to manufacture a product. Hand lay-up requires a low initial investment, but it becomes more expensive (cost per product) than the other techniques due to recurring costs associated with direct labor and material waste. Compared with compression moulding (CM), RTM requests lower tooling costs because of the absence of a press system to compact the preform. RTM seems to meet both low cost/high volume requirements of the automotive industry (500 to 50,000 parts per year) and low number/high performances (50 to 500 part per year) of the aerospace industry. In fact, RTM can guarantee the demanded performances for the aeronautical production: reduction of the mass, increase of the operating life, aimed design, reduction of the production times. Today an increasingly number of parts realized using RTM is observed due to the development of new resins and the preforming technology. In addition, such technique is suitable, with only small adjustments, for the realization of large, complex and thick-walled structures for use in infrastructures and military applications.

Overview of RTM Process

The RTM is a process with a rigid closed mold. Figure below summarizes the main steps for a simple case. The lamination sequence (preform) is draped in a half mold, then the mold is closed and the preform compacted. After that, the resin is injected using a positive gradient pressure through the gate points replacing the air entrapped within the preform. Usually, vacuum is applied at dedicated vents in order to favorite the air escape from the mold. When the resin reaches the vents, the gates are clamped and the preform is impregnated. At this point, the cure phase is considered to start. Finally, the mold is opened and the part removed. Especially for aerospace structures, an additional free-mold post-curing phase can be necessary in order to guarantee the polymerization of the matrix and release the internal thermal stress.

The closing mold step is characterized by the compaction of the fiber reinforcement, which permits to reach the desired thickness and design fiber volume fraction. The compaction changes the microstructure and the dimensions of the preform, producing large deformations and nonlinear viscoelastic effects. These effects are accompanied by a change in energy within the material, which causes the residual stresses due to the viscoelastic behavior of the fibers. However, during the impregnation phase a release of stress, probably due to the balance, occurs.

The injection phase must guarantee the complete impregnation of the preform: a bad impregnation of the fibers results in dry spot areas with missing adhesion between the layers, which makes the surface rough and irregular. If partial impregnation occurs in the proximity of a connecting zone among elements, it can cause a bad integration with a consequent loss of mechanical properties.

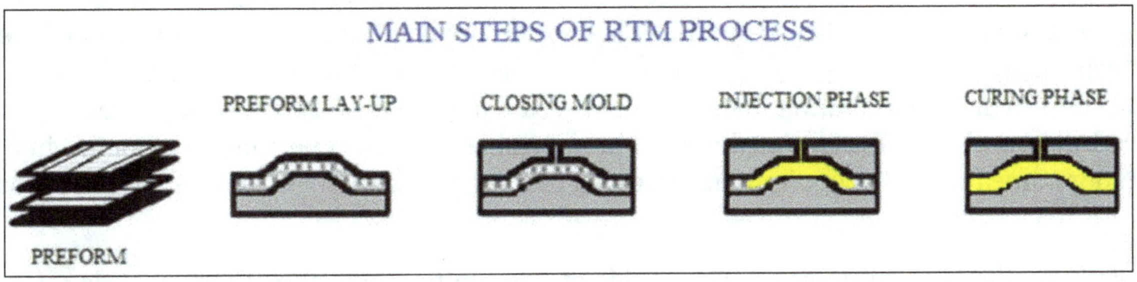

Sequence of the main steps of RTM process. From left to right: realization of the preform, deposition and draping of the preform in the mold cavity, closing mold, injection phase and curing phase; after the curing phase, the mold is opened and manufacred element is demolded.

Process Parameters

The RTM process is governed by variables and parameters that are dependent on each other. Their combination affects the process and the quality of the finished product. Consequently, they need to be carefully determined. The most important parameters, which cannot be neglected in the design, are pressure, temperature, viscosity, permeability, volume fraction, and filling time of the process. There are also a multitude of parameters that must be considered independently, such as the angle of attack of the nozzle, the orientation of the fibers, the paths of flow and shear rates, the stratification. In fact, the resin tends to flow more quickly in the fiber direction, thus the flow dynamic depends mainly on the type of fabric used and the number of overlapped layers. Sometimes it may be necessary to have a certain number of skins, not for structural reasons, but to obtain a homogeneous distribution of the resin. The thickness of the part to be manufactured can also affect the flow progress and the impregnation of the fibers, causing a high percentage of voids and dry spots. The thickness becomes a critical design constraint especially in the case of the inclusion of reinforcements and ribs.

The injection pressure determines the injection velocity of the resin into the mold, the hydraulic pressure and the holding and closing forces of the mold. Consequently, the injection velocity defines the filling time, which should not be too short to ensure an adequate impregnation of the fibers and, at the same time, the filling must be such as to avoid the risk of incurring in premature gelation of the resin. The injection pressure adjusts the distribution of the resin on the preform, which affects the formation of air voids in the matrix, the appearance surface and the mechanical properties of the finished product. Another phenomenon in which this parameter is relevant, together with the viscosity, is the so-called “fiber wash”, i.e. the movement of the reinforcement inside the mold during the injection phase. In this case, the surface treatment of the fibers and especially the choice of the binder play a fundamental role. If the binder dissolves too quickly in contact with the resin, then fibers under the injection pressure can move freely.

The temperature is an extremely important process parameter and it is strictly related to the injection pressure and the viscosity of the resin. When the temperature increases, the filling time decreases and the working pressures are lower. When the temperature is low, the viscosity of the resin increases and it is necessary to increase the pressure to ensure the transfer of the resin itself.

Preform Technology

The diffusion of RTM process in the aerospace industries is strictly related to the possibility of building net-shape 3D-complex structures. This possibility is given by the development of the preform technology. Preform is prepared separately and constitutes the skeleton of the final product, greatly simplifying the moulding operations and reducing the time and cost of processing. When the production volumes is medium-high, an industrialized preform permits to amortize in a relatively short time the cost of equipment, even in a limited production, as required in the aerospace industry.

The textile methods are numerous. The choice of one method over the other depends on several factors: the processability, the feasibility of the geometry, the desired mechanical properties of the molded part, and the cost of production. Obviously, the preform affects strongly the performance required by the final application. The choice of the architecture of the fiber reinforcement depends on the required performance of the composite structure and the characteristics related to the process, such as permeability, compressibility and drape.

The preforms are formed weaving yarns or rovings. The terms used for the fiber comes from the textile tradition: a single fiber is a filament; a set of fibers produced simultaneously is called strand. Several parallel strands can be rolled up like a ribbon, called “roving” or “tows”, or twisted and united as a strand, which is named “yarn”. Generally, the yarns are typically not too complex because the excessive torque reduces the possibility of penetration of the resin in the cavities. Besides, a slight twist of the yarn compacts and raises the formation of composites with high fiber content.

Strands and yarns can be processed as woven roving and cloth. The first type uses effectively the resistance of the fibers, but it can produce composites with high resin content as the roving is not compacted. Woven roving is used to rapidly produce thick composites, because resin can fill them easily. Cloth is slightly less resistant than the roving because of the slight damage that is produced by rolling the fibers and also due to the twisting of the fibers themselves. The cloth is well impregnated from the resin and, as it is compact, high fiber content is obtained.

The weaving process can produce a wide variety of forms. The ratio between the number of filaments in the transverse direction (weft) and the number of filaments in the longitudinal direction (warp) may vary from nearly pure unidirectional, in which the number of filaments of the weft is the minimum required to hold together the fabric, up to the case of equal number of filaments in the warp and weft. In addition, there are hybrid fabrics with carbon fibers in the warp and glass fibers as weft. The weaving method influences the properties of the composite. In flat woven, warp and weft overlap in an alternating manner. In particular, weaving the warp passes through a number of fibers of the weft. These fabrics are better suitable for complex shapes, and their strength and stiffness are slightly higher because the fibers are straight on average.

There are also more complex forms of weaving, with an angle of intersection fibers different from 90°. These methods are used to produce biaxial and multi-axial weaves. The complex fiber architectures can be obtained with the weaving method by interlacing and knitting the fibers along the three spatial directions; an example is shown in Figure below. Different bi-axial layers can be also stitched. The stitching method consists on darning the layers with fibers. These are automated techniques that realize complex shapes and 3-D junctions in place of bolts and rivets. From the mechanical point of view, these methods can increase the resistance to fracture of the structure and reduce the progress of cracks by adopting high strength fibers in the z direction. On the other hand, the stitching seams can produce local defects induced in the preform as a result of penetration of the wire and the needle. Further, the robotic system is very expensive and sometimes damage due to misalignment of fibers can occur.

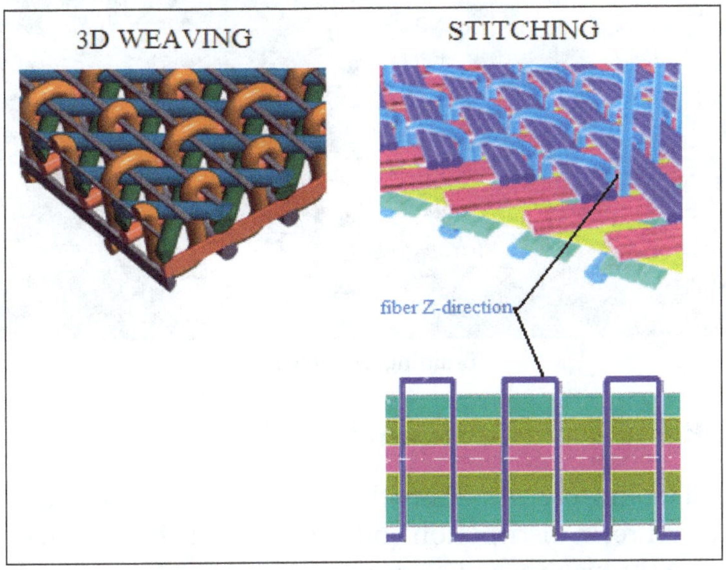

Examples of 3D weaving and stitching textile processes.

The choice of the fiber reinforcement is carried out according to the mechanical requirements of the manufactured component. The types of fibers that can be used are many and with variable characteristics. During machining, the preform can be damaged, particularly when using glass fibers that are fragile and have a high friction coefficient (about 1 in the contact glass on glass). For this reason, immediately after production, the fibers are protected with a coating, "finish" or "size", which performs several functions: it acts as a binder to hold the fibers together, as a lubricant to reduce the coefficient of friction between fibers, and it improves the wettability of the fiber and the adhesion of the resin. Purpose of sizing is also to protect the fibers from the environment, mainly by moisture. These protective sizes are specific to each type of resin, since they must be dissolved from the resin itself. In the case of carbon fibers, the epoxy resin is also used as a protective finish.

Braiding method is another interesting textile process for aerospace structures. The braiding has high level of conformability to complex shapes and it is especially suitable for conical and cylindrical geometry. It regards 2D and 3D preforms. The high level of process automation reduces scrap and labour costs. More advantages of the braiding technique compared to standard tape and fabrics are strength in third dimension, improving fatigue resistance, more efficient distribution of mechanical stresses and the possibility to consider inserts during the preform realization. Despite of these characteristics, the braiding process requires the use of a mandrel and high initial cost for the preform engineering.

An important target of any textile process is obtaining the desired fiber volume fraction that is given by the design. It is very difficult to determine precisely the volume of reinforcement that must match the structural design with the manufacturing aspects. If the fibers are too compacted or their content is excessive, there is no sufficient space for the passage of the resin and the filling time becomes longer. Generally, the optimum volume of fibers is 60% of the product. An increase with respect to this value may cause a bad distribution of the resin and a dramatic drop of the mechanical properties of the manufactured component.

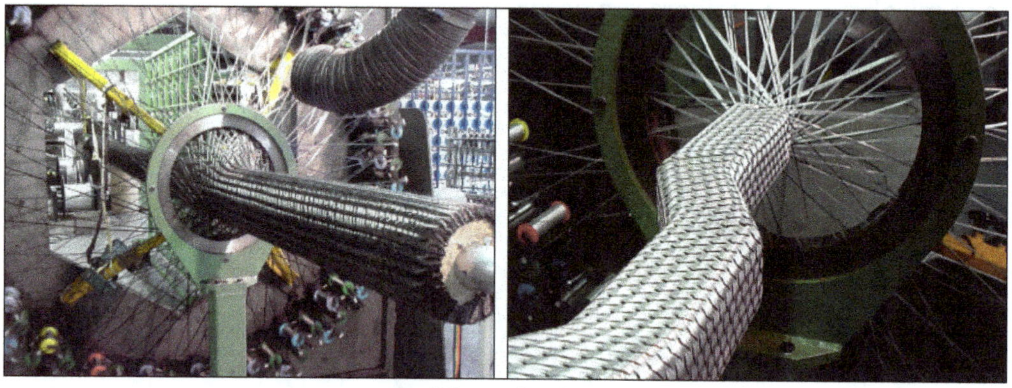

Braiding technology.

Vacuum Assisted Resin Transfer Moulding (VARTM)

VARTM is a single sided tooling process where the dry preform is placed into the tool and vacuum bagged in conjunction with resin distribution and vacuum distribution lines. A low viscosity resin is drawn into the preform through the aid of vacuum. In VARTM process, the flow of resin occurs in plane as well as in the transverse directions to the preform. The permeability of the preform, fiber architecture and fabric crimp has an influence on the wetting of the fabric. This process has certain advantages as listed below:

- Relatively low cost for high volume production.
- Simple low cost tooling.
- Very large and complex parts are practical.
- High fiber volume fraction than hand lay-up.
- On site manufacturing and repairing is possible.
- Reduced environmental concerns than hand lay-up as it is closed system.

This process is being currently used in many of the applications in the general aviation industry, defense sector and in the transport industry.

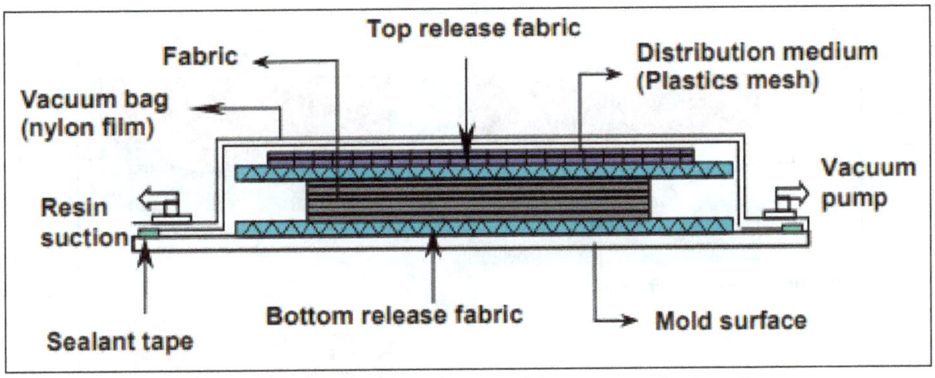

Schematic for the Fabrication of the Composite Panel.

Pultrusion Process

The pultrusion process represents a continuous transportation of fiber bundles through a resin matrix bath, following by a dropping of them into a preheated die or a set of dies. After curing process, where the part is changed from wet saturated reinforcement to a solid par the pultrusion is saw-cut to desired length. A flow diagram of process is showed in Figure below.

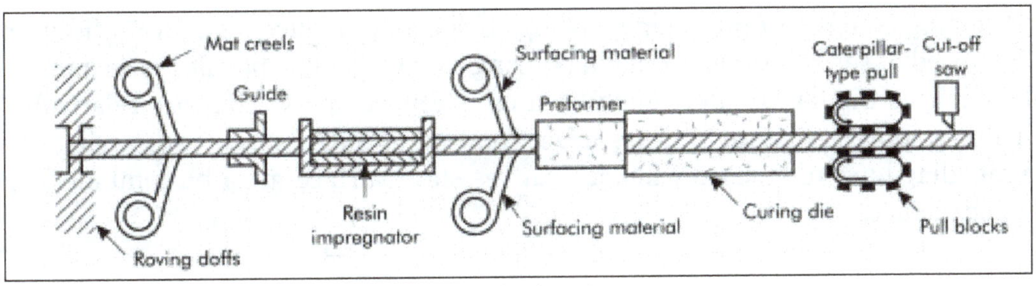

Flow diagram of pultrusion process.

With this process results parts with complex shapes, such as tubing, channels, I-beams, Z-sections and flat bars. For the composites, the pultrusion process is equivalent of metals extrusions, with difference that at pultrution the part is pulled from the exit end of die.

Compression Moulding

It uses preheated molds mounted on a hydraulic or mechanical press. A prepared reinforcement package from prepreg is placed in between the two halves of the mold, which are then pressed against each other to get a desired shape of the mold. Figure below represents the stepwise processing of compression moulding. It offers short cycle time, a high degree of productivity, and automation with dimensional stability, hence it finds diverse applications in the automobile industry. Dispersion of 35% filler elements containing sisal fiber and zirconium dioxide (ZrO_2) particles in the matrix of unsaturated polyester (UP) was obtained by the compression moulding technique, which displayed optimum mechanical properties when tested under SEM, X-ray diffraction, and Fourier transform infrared spectrometer (FTIR). Jute fiber-reinforced epoxy polymer matrix-based composite has been fabricated by using hand lay-up followed by the compression moulding technique at s curing temperature ranging from 80 °C to 130 °C. Enhancement in the mechanical properties has been observed with the maximum tensile strength of 32.3 MPa, flexural strength of 41.8 MPa, and impact strength of 3.5 Joules.

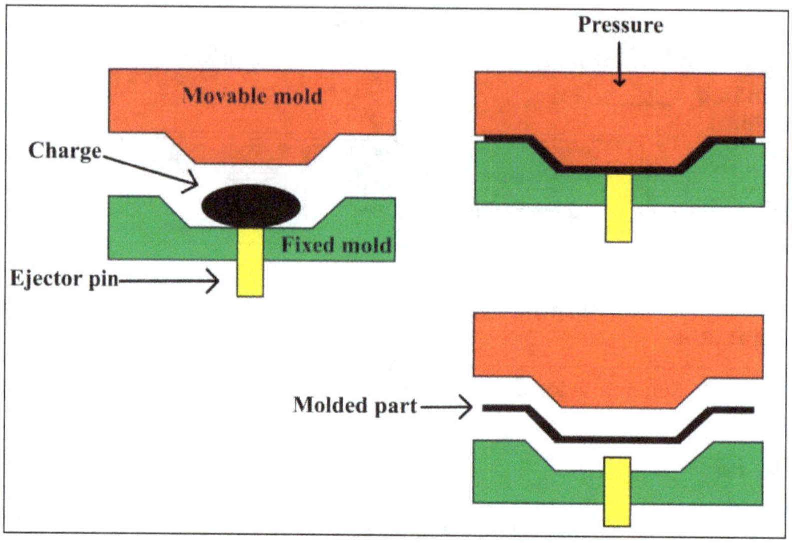

Compression moulding process.

Centrifugal Casting

Centrifugal casting consists of producing castings by causing molten metal to solidify in rotating moulds. The speed of the rotation and metal pouring rate vary with the alloy, size and shape being cast. Centrifugal casting has greater reliability than static castings. They are relatively free from gas and shrinkage porosity. Many times, surface treatments such as case carburizing, flame hardening and nitriding have to be used when a wear resistant surface must be combined with a hard tough exterior surface.

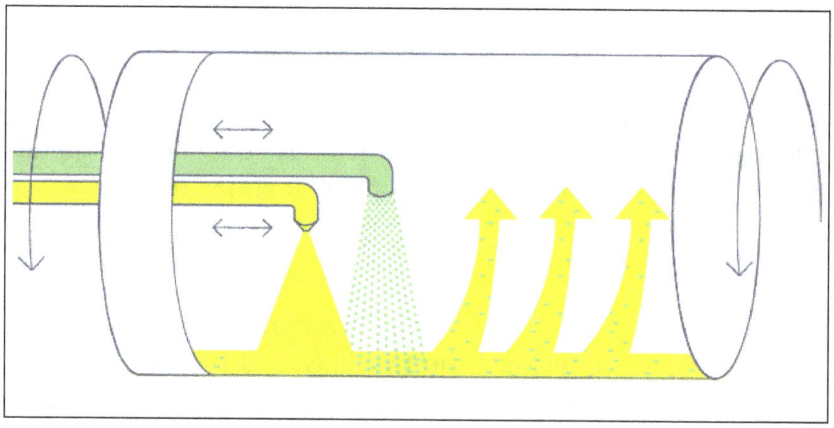

Centrifugal casting consists of a number of processes in which the centrifugal force set up by the rotation of a part of the casting is utilized to shape the casting, fill the mould and help solidify and strengthen the metal. There is a difference between vertical centrifugal casting and horizontal centrifugal casting. The first mentioned process is essentially a pressure casting technique, employing rotation about a vertical axis. It produces good filling of the mould, high dimensional accuracy and a high strength of the cast metal. This method is used for casting of components that are too difficult to produce satisfactorily by static casting methods because their sections are too thin or for other reasons, e.g. gears, piston ring, impellers, propellers, bushings and railway wheels. Horizontal centrifugal casting is used mainly for making long hollow castings, such as pipes, gun barrels, sleeves, rods etc. Centrifugal casting process is industrially used for production of axially symmetrical components such as pipes, cylinder lining and rings.

Continuous Lamination

Continuous lamination is used to produce composite products such as opaque and translucent flat or corrugated paneling, truck trailer paneling, refrigerator liners, sanitary paneling, road signs and other similar products.

Typically, high output machines up to 10 feet (3 meters) wide combine reinforcement and resin on plastic film that is pulled through the process. A second plastic film is applied over the reinforcement and resin to allow mixing and exclusion of air that is usually accomplished by compaction rollers.

Cure is completed in an oven. Panels are automatically trimmed to width and cut to length. Corrugated sheet is produced by forming shoes which hold the compacted sheet in the required shape during cure. Special surface effects are created by using embossed carrier films that are later removed. Both mat reinforcements and rovings chopped by special wide cutters are employed in the process. Polyester and acrylic modified polyesters (for improved water resistance) are the primary resins for continuous lamination.

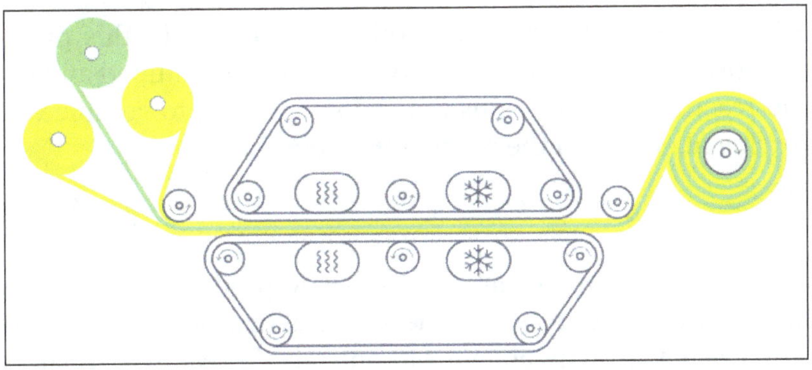

Reinforced Reaction Injection Moulding

In the RRIM process, two or more reactive resins are metered and impingement-mixed under high pressure to form a thermosetting polymer, injected into a mold, and then cured. RRIM composites have a number of processing advantages including very fast cycle time, low labor, low mold clamping pressure and low scrap rate. RRIM uses reinforcements to improve the properties of the resin. With the use of reinforcements, polymerization shrinkage is reduced, thermal expansion is reduced, droop and sag of the composite at elevated temperatures is minimized and other key properties such as stiffness, tensile strength and tensile elongation are generally improved. Milled fibers or flakes can be added directly to the resin before reacting in the mixing head.

Metering is accomplished with high pressure pumps or injection cylinders. Typically, a small mixing chamber is used. The two resin streams enter from opposite sides of mix chamber under high pressure. Mixing occurs from the energy-intensive collision of these two resin streams. Although the streams are mixed at very high pressure, the result is a low viscosity liquid. The low viscosity mixed resin is easily injected into the mold at relatively low pressure, 50 psi (345kPa). Polymerization takes place very quickly within the mold cavity with little or no additional heat required.

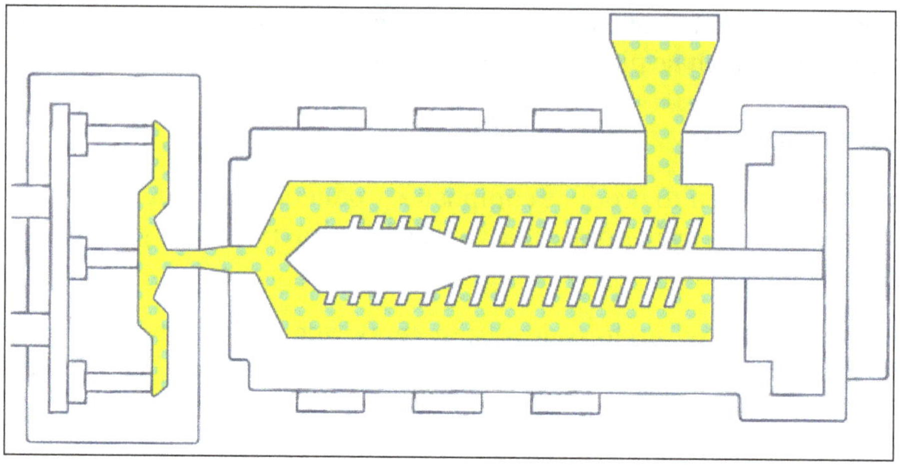

The RRIM process requires special resins and reinforcements. A number of resins, including epoxies, polyesters, nylons and polyurethanes have been successfully developed for RRIM processing. Today, polyurethane is the predominate resin in RRIM. Most of the urethanes used are elastomeric and range in flex modulus from 20,000 psi to well over 500,000 psi (170MPa to 3.5 GPa). The basic RRIM reinforcements are chopped or hammer milled glass fiber and glass flake.

Variations of the RRIM process include structural RIM (SRIM). In this process, chopped fiber preforms or mats are positioned in the mold cavity. The mold is clamped and resin is injected into the mold cavity. The reacting resin remains liquid long enough to completely fill the mold and penetrate the reinforcing fibers. Then the resin quickly cures.

Applications

Presently, transportation is the principal market for RRIM products. Automotive and truck applications for RIIM parts include Class A body panels, fascia, bumper beams, spare tire covers, floor pans and other similar products. The advent of controllable reactivity resins such as polyuria/amide has introduced a trend toward larger machines, larger clamps and larger parts. Very large RRIM molded parts weighing over 100 pounds have already been produced.

Autoclave Moulding

The autoclave consists of a vessel that can be pressurized internally up to 5 bar (~ 75 psi) and then the contents are heated. The vessel must be sufficiently large to accommodate large components. They are pressurized with gas, usually nitrogen that is circulated through the heaters to maintain a uniform temperature throughout the vessel. The basic feedstock for the process is pre-impregnated warp sheets or prepreg. A raw laminate along with a bleeder pack is placed under a nonstick gas permeable film and then that is followed by a breather pack. This whole unit is kept in a vacuum bag to maintain vacuum pressure on the laminate. The outer membrane is pressed against the laminate by atmospheric pressure. The whole unit is then placed in the autoclave where the bagged moulding may be reconnected with the evacuation system to maintain the vacuum. The autoclave is pressurized which augments the consolidated pressure.

The temperature of the autoclave is reduced when the resin is adequately cured. The main aim of this process is to manufacture the laminate with uniform thickness and to ensure minimum porosity. The major difficulty in the autoclave process is the high capitalization cost and the stringent pressure code regulations.

Lanxide Process

The Lanxide process, also known as pressureless metal infiltration, is a way of producing metal-matrix composite materials by a process of partial reaction; the process involves a careful choice of initial alloy (usually aluminium with about 3% magnesium and about 10% silicon), and then the maintenance of conditions in which the polycrystalline reaction product has a mechanical composition such that metal is drawn up through it towards the oxidiser by capillary action, so the composite material grows downwards.

The process can be used for near-net-shape casting, including in protocols where the final casting has a cavity in the shape of the casting pattern - in that case, the metal is poured into the cavity, which has been arranged to lie in the middle of a quantity of "filler" corresponding to the reaction product, and wicks itself to fill the pores in the filler.

The normal application is to produce alumina-reinforced aluminium; the process also allows the growth of ceramic layers inside metal encasements, providing pre-stressing. Lanxide metal-matrix composite materials were used for brake disks in the original models of the Lotus Elise.

Z-Pinning

The Z-pinning technology (also known under Z-fibres reinforcement) consists on the insertion of rigid rods called "pins" in the Z-direction. Z-pins are made with high stiffness and high material strength such as carbon, titanium and steel with a diameter between 0.2-1mm. The Z-pinning process is a simple and reliable method where it requires the placement above the stack of foam containing the pins as shown in figure below. The upper face of the foam is then swept by an ultrasonic impactor which provides the well insertion of pins into the laminate and then it remains to remove the collapsed foam and to cut the protruding. It is necessary to mention that the pins inserted in the foam, act as a fine nails that intertwine the laminate layers together.

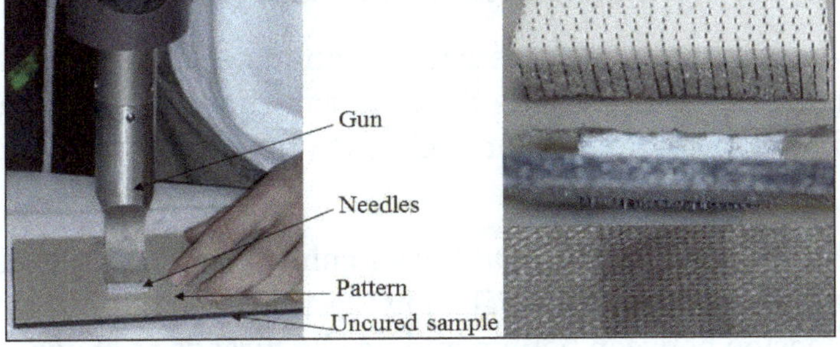

Z-pinning technology.

The most important advantage of Z-pinning technology is the high delamination toughness which increases damage tolerance and impact resistance as well as the joint strength. In addition, Z-fibres process requires just one-sided access to the preform unlike the weaving or stitching process.

Several research have been carried out on the Z-pinning technology shown that this kind of reinforcement improves the interlaminar toughness under different shear modes. Z-pinning technology leads not only to an increase of through-thickness elastic properties of laminates but also to higher impact strength due to its low damage rate.

However, the insertion of pins through laminates thickness may generate a reduction of the in-plane mechanical characteristics. This reduction is due to microstructural damage owing to fibre crimp and waviness which reduces the fiber volume content and obviously the presence of resin-rich zones. Figure below illustrates some microstructure damages caused by pins insertion.

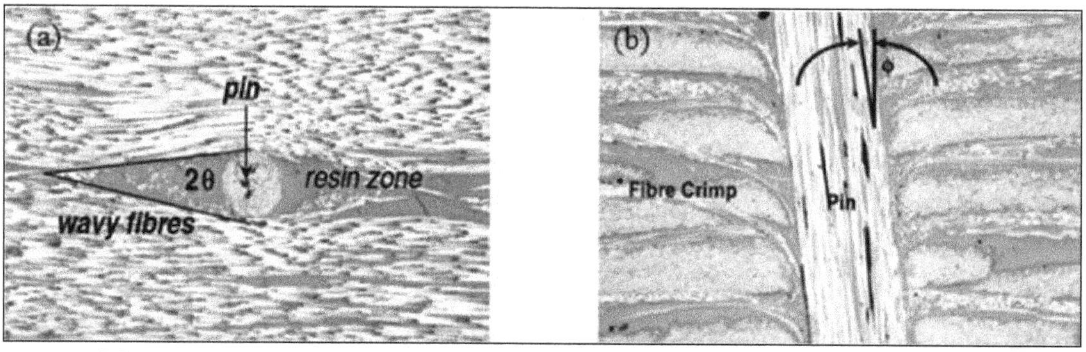

Microstructural damages caused by Z-pinning (a) Fibre waviness and creation of resin-rich zones (b) Fibre crimp.

Stitching Technology

Stitching technology is one of the most effective technology through-the-thickness reinforcement and Dransfield et al. It is based on the sewing process where a dual-threading system (bobbin & needle threads), as shown in figure below (a, b), makes the seam by forming loops and knots through laminates or fabric plies. Stitching involves the insertion of high performance yarns, such as carbon, glass, Kevlar or any other high strength fibrous, through the laminate thickness in order to improve the interlaminar and delamination properties. Typically, there are three different types of stitching techniques:

- Chain stitch: The binding of the loops is carried out at the surface laminate via a single thread. The principle of the chain stitch is summarized in figure below (a).
- Lock stitch: It enquires a double access to the laminate (both bottom and top). The bobbin and needle threads are linked inside the preform which may generates a high stress concentration and resin rich areas. Figure below (b) illustrates the principle of the lock stitch technique.
- Modified lock stitch: It is well known as the stitching type which causes the least damage and stress concentration of the laminate. The intersection is carried out at the top of the preform by varying the tension between bobbin and needle threads as shown in figure below (c).

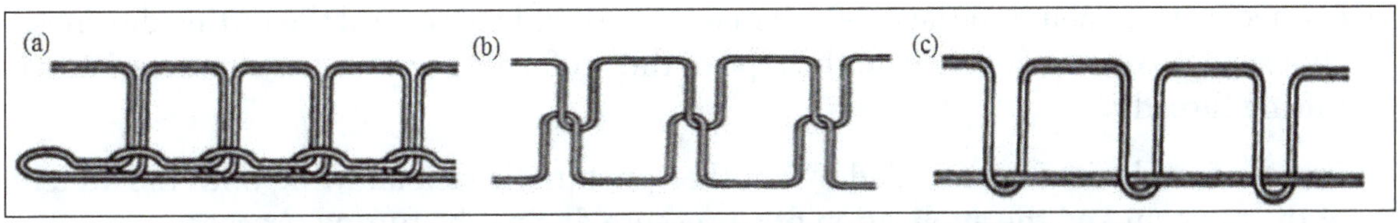

Illustrations of (a) chain stitch (b) lock stitch (c) modified lock stitch.

One of the most effective benefits is that stitch laminates have shown significant improvements in interlaminar fracture toughness, fatigue resistance and impact damage resistance. In the other hand, the main disadvantage of stitching technology is the reduction of the in-plane mechanical performances such as tension, compression and shear properties due to the needle penetration. In particular, the insertion of the needle may generate fibre breakage, misalignment and fibre crimping, hence, a creation of resin-rich pocket within the composite.

Tufting

The tufting technology was used only for carpet and warm garments manufacturing. Now, it has become one of the most efficient technologies for through-the-thickness reinforcement. It is based on the stitching process as shown in figure below. However, this one requires only one-sided access to the preform. The insertion of tufts is carried out via a hollow needle without generating any tension at the surface of the laminate which avoids the shearing and crimping aspects. This kind of reinforcement is more efficient and easier than the stitching technology which requires a second thread as locking thread. In fact, loops are created with simple friction of the tufting thread, which is maintained inside the preform, during the retraction of the needle. Obviously, the tufting process is quiet similar to z-pinning principle, but used to reinforce dry fabrics. Furthermore, the tufting process can be carried out in two different ways. The first one is called 'Global-tufting' where the tufts exceed the depth of the preform and the loops are visible at the bottom of the fabric. The other one is called 'Partial-tufting' where the tufts are fully retained within the laminate. Figure below shows the difference between these categories of tufting. 'Partial tufting' is an efficient technique since it makes it possible to avoid the creation of resin pocket and makes mechanical modeling much simpler.

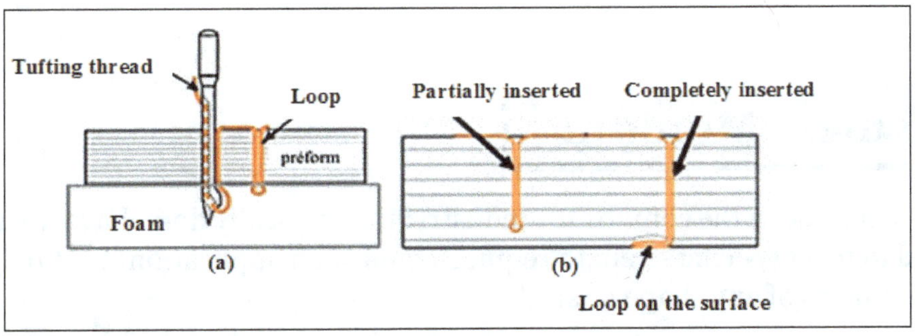

Illustrations of (a) tufting technology (b) partial & global-tufting.

Several works have been carried out in order to automate the tufting device. Figure below presents the automatic equipment developed in our laboratory in order to handle tufting structures.

- Tufting device can move along the x and y axes. It is made with a tufting needle which is connected to a pneumatic jack, which makes it possible to control the tufting deepness. The choice of the needle depends on the type of the preform as well as the characteristics of the tufting thread.

- Presser foot device is controlled by another pneumatic jack allowing both the attachment of the preform and the application of a pressure during the tufting process.
- Feeding device provides the tufting thread with a certain length and tension.
- A mobile framework allowing the movement of the various equipment.

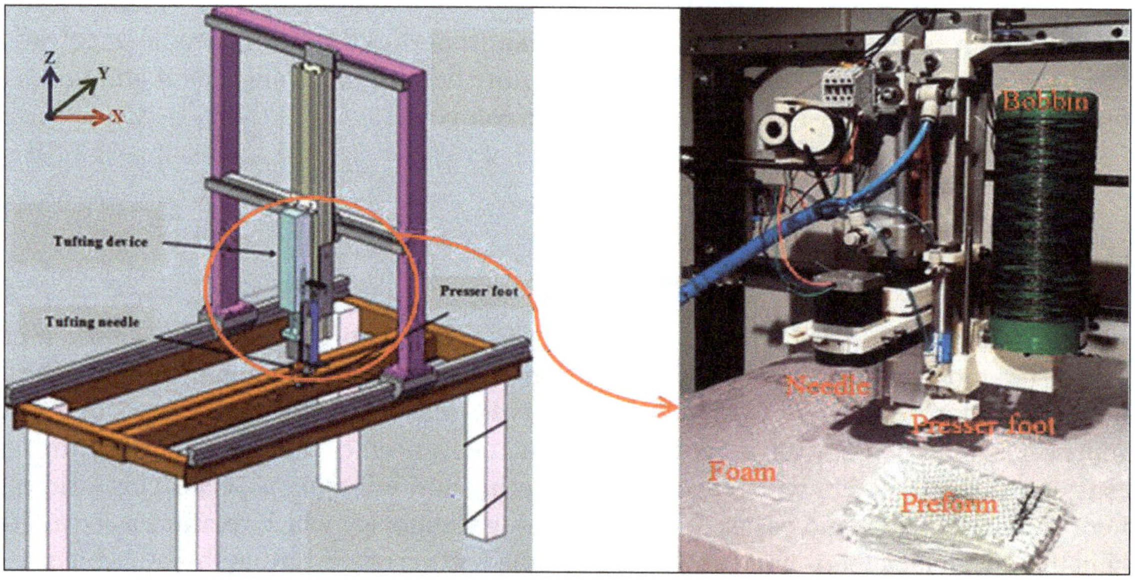

Tufting device (ENSAIT)

In the case of a global-tufting, it should be noted that the choice of the most suitable foam is an essential parameter for the proper handling of the tufting procedure. However, a bad choice of the support generates an irregular formation of the loops and subsequently a complete damage of the loops. Polypropylene or silicone foams are the most widely used. There are several parameters which can influence the final structure such as tufting density, loop length, tufting direction, tufting deepness and pressure of the presser foot.

Cast Polymers Moulding

Cast polymers are unique in the composites industry: they typically don't have fiber reinforcement and are designed to meet specific strength requirements of an application. Cast polymer moulding is used to produce parts of any shape or size.

Gel Coated Cultured Stone

Gel coat is a specialized polyester resin that is formulated to provide a cosmetic outer surface on a composite product and to provide weatherability for outdoor products. Several variations of cultured stone products are manufactured using a gel coated surface and a resin-matrix casting process. In this process a gel coat film (usually clear) is sprayed on the mold surface. Once the gel coat is sufficiently cured, a polyester resin matrix is blended by adding various types of fillers to the resin. Pigments for both a solid background color and the look of veins found in natural stone can be added. The resin matrix is then transferred to the mold, where vibration is applied to level and compact the matrix. Following the cure, the part is removed from the mould.

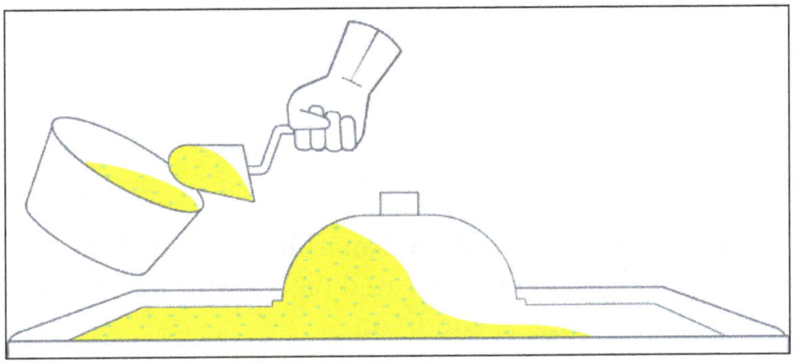

Gel coat is not paint. Paint contains solvents that must evaporate for the paint to dry. The 'solvent' in gel coat is styrene monomer and/or methylmethacrylate (acrylic), which cross-links during curing. The monomer does not have to leave the system for the gel coat to cure; in fact, it is beneficial to reduce monomer loss.

The appearance of the cultured stone products is determined by the type of filler used and by the application of colorants to the matrix. Fillers come in a variety of materials. Many of the fillers used in the composites industry are mineral substances. Mineral fillers have distinctive shapes that relate to their chemical structure.

- Marble: The natural marble look is reproduced by formulating a matrix using calcium carbonate filler. In some cases, other fillers or combinations of fillers may be used. Resin, initiator, filler, and pigment are mixed to form a solid-color matrix. The marble veining effect is created by adding a second pigment to the matrix and partially mixing it to produce the desired look.
- Onyx: The process of manufacturing cultured onyx is similar to that of cultured marble, except alumina trihydrate (ATH) filler is used. The cultured onyx matrix generally has a higher resin content compared with cultured marble, and the combination of materials creates a translucent appearance. Background and veining pigments are added to the matrix to produce an onyx stone look.
- Granite: The cultured granite appearance is created by blending colored chips into the resin matrix. These chips can be made from cultured marble castings, thermoplastics, or even actual granite stone that has been ground into particles. The cultured granite matrix usually consists of polyester resin, initiator, colored chips, and ATH filler.

Solid Surface Moulding

Solid surface is a void-free casting made from a blend of polyester resin or acrylic resin, initiator, ATH, color chips, and pigment. Solid surface can be formulated to achieve a wide variety of looks and cosmetic effects, such as simulating natural granite stone. For this reason, it is often used to manufacture products like kitchen countertops.

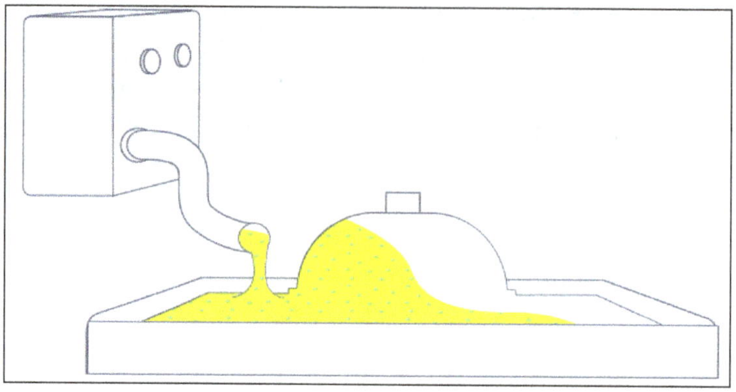

In contrast to the gel coated surface of cultured marble, solid surface parts are homogeneous throughout. This makes it possible to join fabricated pieces with inconspicuous seams and to repair and refinish the surface to its original condition.

Solid surface, or densified, castings are made using vacuum-mixing techniques to produce a matrix that is void free. This produces a material that presents a uniform surface when it is cut, sanded, or bonded. Solid surface castings are post-cured at elevated temperatures (in the range of 200 degrees Fahrenheit) to enhance the physical properties of the matrix and produce a stable product. Solid surface can be compression molded, which is a high-pressure, closed-moulding process suitable for moulding high volumes of complex solid surface parts. The compression moulding process uses matched metal-heated molds mounted in large hydraulic presses.

Compression moulding produces fast moulding cycles and high part uniformity but requires a high capital investment in tooling and equipment. Features such as inserts, ribs, bosses or attachments can be molded in. Good surface finishes can contribute to lower part-finishing costs. Subsequent trimming and machining operations are minimized in compression moulding. This process is capital intensive and labor-efficient. Labor costs are low due to the fast cycle times and reduced post-mold finishing, while capital costs are high for heated-metal tooling and moulding presses.

References

- Composites-manufacturing-processes, spray-up, open-molding: compositeslab.com, Retrieved 20, March 2020
- Pressure-bag-moulding: blogspot.com, Retrieved 10, July 2020
- InTech-Advanced-composite-materials-by-resin-transfer-molding-for-aerospace-applications-38405: intechopen.com, Retrieved 18, April 2020
- Introduction-to-low-cost-manufacturing-of-composite-laminates: asee.org, Retrieved 04, January 2020
- Manufacturing-process-and-applications-of-composite-materials-228766301: researchgate.net, Retrieved 25, August 2020
- Design-and-fabrication-of-a-centrifugal-casting-machine-266055018: researchgate.net, Retrieved 19, March 2020
- Centrifugal-casting, composites-manufacturing-processes: compositeslab.com, Retrieved 06, June 2020
- Gel-coated-cultured-stone-molding, cast-polymer-molding, composites-manufacturing-processes: compositeslab.com, Retrieved 17, April 2020

4

Nanocomposites and its Types

A multiphase solid material wherein at least one phase has a dimension of less than 100 nanometers is known as a nanocomposite. A few of the major types of nanocomposites are polymer nanocomposite, carbon nanotube, nanowire and nanocellulose. This chapter discusses in detail these types of nanocomposites.

Nanocomposite is a multiphase solid material where one of the phases has one, two or three dimensions of less than 100 nanometers (nm), or structures having nano-scale repeat distances between the different phases that make up the material. In the broadest sense this definition can include porous media, colloids, gels and copolymers, but is more usually taken to mean the solid combination of a bulk matrix and nano-dimensional phase(s) differing in properties due to dissimilarities in structure and chemistry. The mechanical, electrical, thermal, optical, electrochemical, catalytic properties of the nanocomposite will differ markedly from that of the component materials. Size limits for these effects have been proposed, <5 nm for catalytic activity, <20 nm for making a hard magnetic material soft, <50 nm for refractive index changes, and <100 nm for achieving superparamagnetism, mechanical strengthening or restricting matrix dislocation movement.

Nanocomposites are found in nature, for example in the structure of the abalone shell and bone. The use of nanoparticle-rich materials long predates the understanding of the physical and chemical nature of these materials. Jose-Yacaman *et al.* investigated the origin of the depth of colour and the resistance to acids and bio-corrosion of Maya blue paint, attributing it to a nanoparticle mechanism. From the mid-1950s nanoscale organo-clays have been used to control flow of polymer solutions (e.g. as paint viscosifiers) or the constitution of gels (e.g. as a thickening substance in cosmetics, keeping the preparations in homogeneous form). By the 1970s polymer/clay composites were the topic of textbooks, although the term “nanocomposites” was not in common use.

In mechanical terms, nanocomposites differ from conventional composite materials due to the exceptionally high surface to volume ratio of the reinforcing phase and/or its exceptionally high aspect ratio. The reinforcing material can be made up of particles (e.g. minerals), sheets (e.g. exfoliated clay stacks) or fibres (e.g. carbon nanotubes or electrospun fibres). The area of the interface between the matrix and reinforcement phase(s) is typically an order of magnitude greater than for conventional composite materials. The matrix material properties are significantly affected in the vicinity of the reinforcement. Ajayan *et al.* note that with polymer nanocomposites, properties related to local chemistry, degree of thermoset cure, polymer chain mobility, polymer chain conformation, degree of polymer chain ordering or crystallinity can all vary significantly and continuously from the interface with the reinforcement into the bulk of the matrix.

This large amount of reinforcement surface area means that a relatively small amount of nanoscale reinforcement can have an observable effect on the macroscale properties of the composite. For example, adding carbon nanotubes improves the electrical and thermal conductivity. Other kinds of nanoparticulates may result in enhanced optical properties, dielectric properties, heat resistance or mechanical properties such as stiffness, strength and resistance to wear and damage.

In general, the nano reinforcement is dispersed into the matrix during processing. The percentage by weight (called *mass fraction*) of the nanoparticulates introduced can remain very low (on the order of 0.5% to 5%) due to the low filler percolation threshold, especially for the most commonly used non-spherical, high aspect ratio fillers (e.g. nanometer-thin platelets, such as clays, or nanometer-diameter cylinders, such as carbon nanotubes). The orientation and arrangement of asymmetric nanoparticles, thermal property mismatch at the interface, interface density per unit volume of nanocomposite, and polydispersity of nanoparticles significantly affect the effective thermal conductivity of nanocomposites.

Ceramic-matrix Nanocomposites

In this group of composites the main part of the volume is occupied by a ceramic, i.e. a chemical compound from the group of oxides, nitrides, borides, silicides etc.. In most cases, ceramic-matrix nanocomposites encompass a metal as the second component. Ideally both components, the metallic one and the ceramic one, are finely dispersed in each other in order to elicit the particular nanoscopic properties. Nanocomposite from these combinations were demonstrated in improving their optical, electrical and magnetic properties as well as tribological, corrosion-resistance and other protective properties.

The binary phase diagram of the mixture should be considered in designing ceramic-metal nanocomposites and measures have to be taken to avoid a chemical reaction between both components. The last point mainly is of importance for the metallic component that may easily react with the ceramic and thereby lose its metallic character. This is not an easily obeyed constraint, because the preparation of the ceramic component generally requires high process temperatures. The most safe measure thus is to carefully choose immiscible metal and ceramic phases. A good example for such a combination is represented by the ceramic-metal composite of TiO_2 and Cu, the mixtures of which were found immiscible over large areas in the Gibbs' triangle of Cu-O-Ti.

The concept of ceramic-matrix nanocomposites was also applied to thin films that are solid layers of a few nm to some tens of μm thickness deposited upon an underlying substrate and that play an important role in the functionalization of technical surfaces. Gas flow sputtering by the hollow cathode technique turned out as a rather effective technique for the preparation of nanocomposite layers. The process operates as a vacuum-based deposition technique and is associated with high deposition rates up to some μm/s and the growth of nanoparticles in the gas phase. Nanocomposite layers in the ceramics range of composition were prepared from TiO_2 and Cu by the hollow cathode technique that showed a high mechanical hardness, small coefficients of friction and a high resistance to corrosion.

Metal-matrix Nanocomposites

Metal matrix nanocomposites can also be defined as reinforced metal matrix composites. This type of composites can be classified as continuous and non-continuous reinforced materials. One of the more important nanocomposites is Carbon nanotube metal matrix composites, which is an emerging new material that is being developed to take advantage of the high tensile strength and electrical conductivity of carbon nanotube materials. Critical to the realization of CNT-MMC possessing optimal properties in these areas are the development of synthetic techniques that are

(a) economically producible, (b) provide for a homogeneous dispersion of nanotubes in the metallic matrix, and (c) lead to strong interfacial adhesion between the metallic matrix and the carbon nanotubes. In addition to carbon nanotube metal matrix composites, boron nitride reinforced metal matrix composites and carbon nitride metal matrix composites are the new research areas on metal matrix nanocomposites.

A recent study, comparing the mechanical properties (Young's modulus, compressive yield strength, flexural modulus and flexural yield strength) of single- and multi-walled reinforced polymeric (polypropylene fumarate—PPF) nanocomposites to tungsten disulfide nanotubes reinforced PPF nanocomposites suggest that tungsten disulfide nanotubes reinforced PPF nanocomposites possess significantly higher mechanical properties and tungsten disulfide nanotubes are better reinforcing agents than carbon nanotubes. Increases in the mechanical properties can be attributed to a uniform dispersion of inorganic nanotubes in the polymer matrix (compared to carbon nanotubes that exist as micron sized aggregates) and increased crosslinking density of the polymer in the presence of tungsten disulfide nanotubes (increase in crosslinking density leads to an increase in the mechanical properties). These results suggest that inorganic nanomaterials, in general, may be better reinforcing agents compared to carbon nanotubes.

Another kind of nanocomposite is the energetic nanocomposite, generally as a hybrid sol–gel with a silica base, which, when combined with metal oxides and nano-scale aluminum powder, can form *superthermite* materials.

Polymer-matrix Nanocomposites

In the simplest case, appropriately adding nanoparticulates to a polymer matrix can enhance its performance, often dramatically, by simply capitalizing on the nature and properties of the nanoscale filler (these materials are better described by the term *nanofilled polymer composites*). This strategy is particularly effective in yielding high performance composites, when good dispersion of the filler is achieved and the properties of the nanoscale filler are substantially different or better than those of the matrix.

Nanoparticles such as graphene, carbon nanotubes, molybdenum disulfide and tungsten disulfide are being used as reinforcing agents to fabricate mechanically strong biodegradable polymeric nanocomposites for bone tissue engineering applications. The addition of these nanoparticles in the polymer matrix at low concentrations (~0.2 weight %) cause significant improvements in the compressive and flexural mechanical properties of polymeric nanocomposites. Potentially, these nanocomposites may be used as a novel, mechanically strong, light weight composite as bone implants. The results suggest that mechanical reinforcement is dependent on the nanostructure morphology, defects, dispersion of nanomaterials in the polymer matrix, and the cross-linking density of the polymer. In general, two-dimensional nanostructures can reinforce the polymer better than one-dimensional nanostructures, and inorganic nanomaterials are better reinforcing agents than carbon based nanomaterials. In addition to mechanical properties, polymer nanocomposites based on carbon nanotubes or graphene have been used to enhance a wide range of properties, giving rise to functional materials for a wide range of high added value applications in fields such as energy conversion and storage, sensing and biomedical tissue engineering. For example, multi-walled carbon nanotubes based polymer nanocomposites have been used for the enhancement of the electrical conductivity.

Nanoscale dispersion of filler or controlled nanostructures in the composite can introduce new physical properties and novel behaviors that are absent in the unfilled matrices. This effectively changes the nature of the original matrix (such composite materials can be better described by the term *genuine nanocomposites or hybrids*). Some examples of such new properties are fire resistance or flame retardancy, and accelerated biodegradability.

A range of polymeric nanocomposites are used for biomedical applications such as tissue engineering, drug delivery, cellular therapies. Due to unique interactions between polymer and nanoparticles, a range of property combinations can be engineered to mimic native tissue structure and properties. A range of natural and synthetic polymers are used to design polymeric nanocomposites for biomedical applications including starch, cellulose, alginate, chitosan, collagen, gelatin, and fibrin, poly(vinyl alcohol) (PVA), poly(ethylene glycol) (PEG), poly(caprolactone) (PCL), poly(lactic-co-glycolic acid) (PLGA), and poly(glycerol sebacate) (PGS). A range of nanoparticles including ceramic, polymeric, metal oxide and carbon-based nanomaterials are incorporated within polymeric network to obtain desired property combinations.

Polymer Nanocomposite

Polymer nanocomposites (PNC) consist of a polymer or copolymer having nanoparticles or nanofillers dispersed in the polymer matrix. These may be of different shape (e.g., platelets, fibers, spheroids), but at least one dimension must be in the range of 1–50 nm. These PNC's belong to the category of multi-phase systems (MPS, viz. blends, composites, and foams) that consume nearly 95% of plastics production. These systems require controlled mixing/compounding, stabilization of the achieved dispersion, orientation of the dispersed phase, and the compounding strategies for all MPS, including PNC, are similar.

Polymer nanoscience is the study and application of nanoscience to polymer-nanoparticle matrices, where nanoparticles are those with at least one dimension of less than 100 nm.

The transition from micro- to nano-particles lead to change in its physical as well as chemical properties. Two of the major factors in this are the increase in the ratio of the surface area to volume, and the size of the particle. The increase in surface area-to-volume ratio, which increases as the particles get smaller, leads to an increasing dominance of the behavior of atoms on the surface area of particle over that of those interior of the particle. This affects the properties of the particles when they are reacting with other particles. Because of the higher surface area of the nano-particles, the interaction with the other particles within the mixture is more and this increases the strength, heat resistance, etc. and many factors do change for the mixture.

An example of a nanopolymer is silicon nanospheres which show quite different characteristics; their size is 40–100 nm and they are much harder than silicon, their hardness being between that of sapphire and diamond.

Bio-hybrid Polymer Nanofibers

Many technical applications of biological objects like proteins, viruses or bacteria such as chromatography, optical information technology, sensorics, catalysis and drug delivery require their immobilization. Carbon nanotubes, gold particles and synthetic polymers are used for this purpose. This immobilization has been achieved predominantly by adsorption or by chemical binding and to a lesser extent by incorporating these objects as guests in host matrices. In the guest host systems, an ideal method for the immobilization of biological objects and their integration into hierarchical architectures should be structured on a nanoscale to facilitate the interactions of biological nano-objects with their environment. Due to the large number of natural or synthetic polymers available and the advanced techniques developed to process such systems to nanofibres, rods, tubes etc. make polymers a good platform for the immobilization of biological objects.

Bio-Hybrid Nanofibres by Electrospinning

Polymer fibers are, in general, produced on a technical scale by extrusion, i.e., a polymer melt or a polymer solution is pumped through cylindrical dies and spun/drawn by a take-up device. The resulting fibers have diameters typically on the 10-µm scale or above. To come down in diameter into the range of several hundreds of nanometers or even down to a few nanometers, Electrospinning is today still the leading polymer processing technique available. A strong electric field of the order of 103 V/cm is applied to the polymer solution droplets emerging from a cylindrical die. The electric charges, which are accumulated on the surface of the droplet, cause droplet deformation along the field direction, even though the surface tension counteracts droplet evolution. In supercritical electric fields, the field strength overbears the surface tension and a fluid jet emanates from the droplet tip. The jet is accelerated towards the counter electrode. During this transport phase, the jet is subjected to strong electrically driven circular bending motions that cause a strong elongation and thinning of the jet, a solvent evaporation until, finally, the solid nanofibre is deposited on the counter electrode.

Bio-Hybrid Polymer Nanotubes by Wetting

Electro spinning, co-electrospinning, and the template methods based on nanofibres yield nano-objects which are, in principle, infinitively long. For a broad range of applications including catalysis, tissue engineering, and surface modification of implants this infinite length is an advantage. But in some applications like inhalation therapy or systemic drug delivery, a well-defined length is required. The template method to be described in the following has the advantage such that it allows the preparation of nanotubes and nanorods with very high precision. The method is based on the use of well defined porous templates, such as porous aluminum or silicon.

The basic concept of this method is to exploit wetting processes. A polymer melt or solution is brought into contact with the pores located in materials characterized by high energy surfaces such as aluminum or silicon. Wetting sets in and covers the walls of the pores with a thin film with a thickness of the order of a few tens of nanometers.

Gravity does not play a role, as it is obvious from the fact that wetting takes place independent of the orientation of the pores relative to the direction of gravity. The exact process is still not understood theoretically in detail but its known from experiments that low molar mass systems tend to fill the pores completely, whereas polymers of sufficient chain length just cover the walls. This process happens typically within a minute for temperatures about 50 K above the melting temperature or glass transition temperature, even for highly viscous polymers, such as, for instance, polytetrafluoroethylene, and this holds even for pores with an aspect ratio as large as 10,000. The complete filling, on the other hand, takes days. To obtain nanotubes, the polymer/template system is cooled down to room temperature or the solvent is evaporated, yielding pores covered with solid layers. The resulting tubes can be removed by mechanical forces for tubes up to 10 μm in length, i.e., by just drawing them out from the pores or by selectively dissolving the template. The diameter of the nanotubes, the distribution of the diameter, the homogeneity along the tubes, and the lengths can be controlled.

Applications

The nanofibres, hollow nanofibres, core–shell nanofibres, and nanorods or nanotubes produced have a great potential for a broad range of applications including homogeneous and heterogeneous catalysis, sensorics, filter applications, and optoelectronics. Here we will just consider a limited set of applications related to life science.

Tissue Engineering

This is mainly concerned with the replacement of tissues which have been destroyed by sickness or accidents or other artificial means. The examples are skin, bone, cartilage, blood vessels and may be even organs. This technique involves providing a scaffold on which cells are added and the scaffold should provide favorable conditions for the growth of the same. Nanofibres have been found to provide very good conditions for the growth of such cells, one of the reasons being that fibrillar structures can be found on many tissues which allow the cells to attach strongly to the fibers and grow along them as shown.

Nanoparticles such as graphene, carbon nanotubes, molybdenum disulfide and tungsten disulfide are being used as reinforcing agents to fabricate mechanically strong biodegradable polymeric nanocomposites for bone tissue engineering applications. The addition of these nanoparticles in the polymer matrix at low concentrations (~0.2 weight %) leads to significant improvements in the compressive and flexural mechanical properties of polymeric nanocomposites. Potentially, these nanocomposites may be used as a novel, mechanically strong, light weight composite as bone implants. The results suggest that mechanical reinforcement is dependent on the nanostructure morphology, defects, dispersion of nanomaterials in the polymer matrix, and the cross-linking density of the polymer. In general, two-dimensional nanostructures can reinforce the polymer better than one-dimensional nanostructures, and inorganic nanomaterials are better reinforcing agents than carbon based nanomaterials.

Delivery from Compartmented Nanotubes

Nano tubes are also used for carrying drugs in general therapy and in tumor therapy in particular. The role of them is to protect the drugs from destruction in blood stream, to control the delivery with a well-defined release kinetics, and in ideal cases, to provide vector-targeting properties or release mechanism by external or internal stimuli.

Rod or tube-like, rather than nearly spherical, nanocarriers may offer additional advantages in terms of drug delivery systems. Such drug carrier particles possess additional choice of the axial ratio, the curvature, and the "all-sweeping" hydrodynamic-related rotation, and they can be modified chemically at the inner surface, the outer surface, and at the end planes in a very selective way. Nanotubes prepared with a responsive polymer attached to the tube opening allow the control of access to and release from the tube. Furthermore, nanotubes can also be prepared showing a gradient in its chemical composition along the length of the tube.

Compartmented drug release systems were prepared based on nanotubes or nanofibres. Nanotubes and nanofibres, for instance, which contained fluorescent albumin with dog-fluorescein isothiocyanate were prepared as a model drug, as well as super paramagnetic nanoparticles composed of iron oxide or nickel ferrite. The presence of the magnetic nanoparticles allowed, first of all, the guiding of the nanotubes to specific locations in the body by external magnetic fields. Super paramagnetic particles are known to display strong interactions with external magnetic fields leading to large saturation magnetizations. In addition, by using periodically varying magnetic fields, the nanoparticles were heated up to provide, thus, a trigger for drug release. The presence of the model drug was established by fluorescence spectroscopy and the same holds for the analysis of the model drug released from the nanotubes.

Immobilization of Proteins

Core shell fibers of nano particles with fluid cores and solid shells can be used to entrap biological objects such as proteins, viruses or bacteria in conditions which do not affect their functions. This effect can be used among others for biosensor applications. For example, Green Fluorescent Protein is immobilized in nanostructured fibres providing large surface areas and short distances for the analyte to approach the sensor protein.

With respect to using such fibers for sensor applications fluorescence of the core shell fibers was found to decay rapidly as the fibers were immersed into a solution containing urea: urea permeates through the wall into the core where it causes denaturation of the GFP. This simple experiment reveals that core–shell fibers are promising objects for preparing biosensors based on biological objects.

Polymer nanostructured fibers, core–shell fibers, hollow fibers, and nanorods and nanotubes provide a platform for a broad range of applications both in material science as well as in life science. Biological objects of different complexity and synthetic objects carrying specific functions can be incorporated into such nanostructured polymer systems while keeping their specific functions vital. Biosensors, tissue engineering, drug delivery, or enzymatic catalysis is just a few of the possible examples. The incorporation of viruses and bacteria all the way up to microorganism should not really pose a problem and the applications coming from such biohybrid systems should be tremendous.

Size and Pressure Effects on Nanopolymers

The size- and pressure- dependent glass transition temperatures of free-standing films or supported films having weak interactions with **substrates** decreases with decreasing of pressure and size. However, the glass transition temperature of supported films having strong interaction with substrates increases of pressure and the decrease of size. Different models like two layer model, three layer model, T_g (D, 0) 1/D and some more models relating specific heat, density and thermal expansion are used to obtain the experimental results on nanopolymers and even some observations like freezing of films due to memory effects in the visco-elastic eigenmodels of the films, and finite effects of the small molecule glass are observed. To describe T_g (D, 0) function of polymers more generally, a simple and unified model recently is provided based on the size-dependent melting temperature of crystals and Lindemann's criterion,

$$T_g(D, 0) / T_g(\infty, 0)\, \sigma_g^2(\infty, 0) / \sigma_g^2(D, 0)$$

where σ_g is the root of mean squared displacement of surface and interior molecules of glasses at T_g (D, 0), $\alpha = \sigma_s^2$ (D, 0) / σ_v^2 (D, 0) with subscripts *s* and *v* denoting surface and volume, respectively. For a nanoparticle, D has a usual meaning of diameter, for a nanowire, D is taken as its diameter, and for a thin film, D denotes its thickness. D_0 denotes a critical diameter at which all molecules of a low-dimensional glass are located on its surface.

Conclusion

The devices that use the properties of low-dimensional objects such as nanoparticles are promising due to the possibility of tailoring a number of electrophysical, optical and magnetic properties changing the size of nanoparticles, which can be controlled during the synthesis. In the case of polymer nanocomposites we can use the properties of disordered systems.

Here recent developments in the field of polymer nano-composites and some of their applications have been reviewed. Though there is much use in this field, there are many limitations also. For example, in the release of drugs using nanofibres, cannot be controlled independently and a burst release is usually the case, whereas a more linear release is required. Let us now consider future aspects in this field.

There is a possibility of building ordered arrays of nanoparticles in the polymer matrix. A number of possibilities also exist to manufacture the nanocomposite circuit boards. An even more attractive method exists to use polymer nanocomposites for neural networks applications. Another promising area of development is optoelectronics and optical computing. The single domain nature and super paramagnetic behavior of nanoparticles containing ferromagnetic metals could be possibly used for magneto-optical storage media manufacturing.

Carbon Nanotube

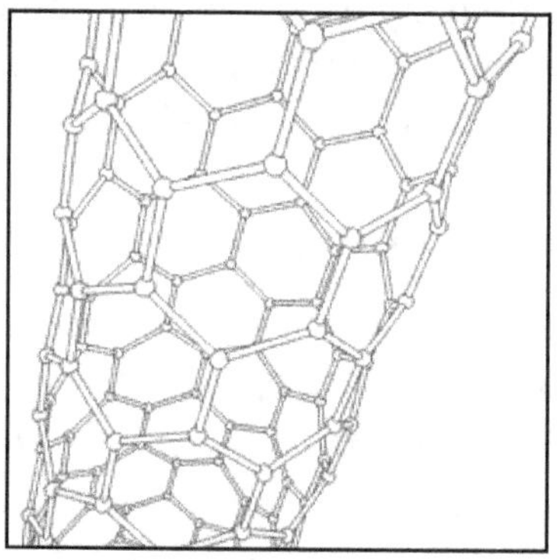

Rotating single-walled zigzag carbon nanotube.

Carbon nanotubes (CNTs) are allotropes of carbon with a cylindrical nanostructure. Nanotubes have been constructed with length-to-diameter ratio of up to 132,000,000:1, significantly larger than for any other material. These cylindrical carbon molecules have unusual properties, which are valuable for nanotechnology, electronics, optics and other fields of materials science and technology. In particular, owing to their extraordinary thermal conductivity and mechanical and electrical properties, carbon nanotubes find applications as additives to various structural materials.

For instance, nanotubes form a tiny portion of the material(s) in some (primarily carbon fiber) baseball bats, golf clubs, car parts or damascus steel.

Nanotubes are members of the fullerene structural family. Their name is derived from their long, hollow structure with the walls formed by one-atom-thick sheets of carbon, called graphene. These sheets are rolled at specific and discrete ("chiral") angles, and the combination of the rolling angle and radius decides the nanotube properties; for example, whether the individual nanotube shell is a metal or semiconductor. Nanotubes are categorized as single-walled nanotubes (SWNTs) and multi-walled nanotubes (MWNTs). Individual nanotubes naturally align themselves into "ropes" held together by van der Waals forces, more specifically, pi-stacking.

Applied quantum chemistry, specifically, orbital hybridization best describes chemical bonding in nanotubes. The chemical bonding of nanotubes is composed entirely of sp^2 bonds, similar to those of graphite. These bonds, which are stronger than the sp^3 bonds found in alkanes and diamond, provide nanotubes with their unique strength.

Types of Carbon Nanotubes and Related Structures

Terminology

There is no consensus on some terms describing carbon nanotubes in scientific literature: both "-wall" and "-walled" are being used in combination with "single", "double", "triple" or "multi", and the letter C is often omitted in the abbreviation; for example, multi-walled carbon nanotube (MWNT).

Single-walled

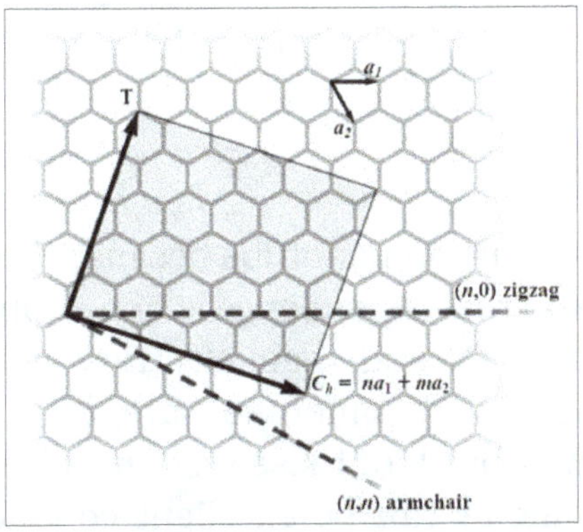

The (n,m) nanotube naming scheme can be thought of as a vector ($\mathbf{C}_h$) in an infinite graphene sheet that describes how.

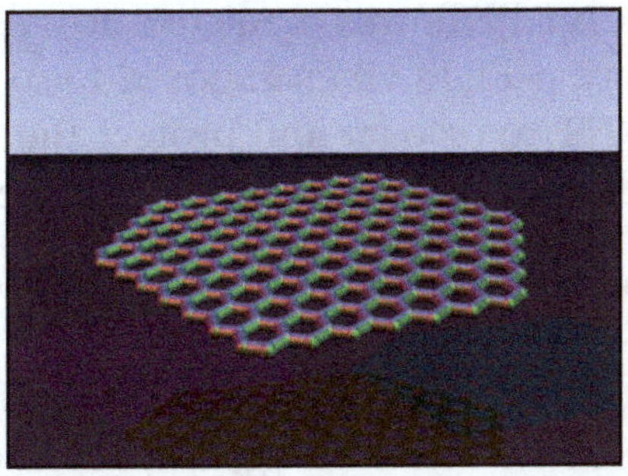

Graphene nanoribbon to "roll up" the graphene sheet to make the nanotube. **T** denotes the tube axis, and $\mathbf{a}_1$ and $\mathbf{a}_2$ are the unit vectors of graphene in real space.

A scanning tunneling microscopy image of single-walled carbon nanotube.

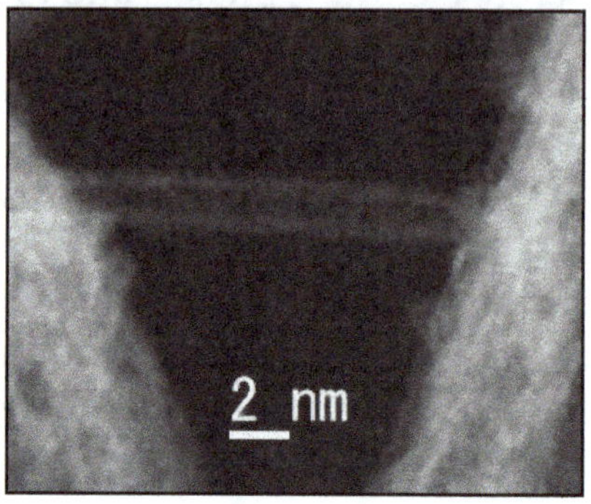

A transmission electron microscopy image of a single-walled carbon nanotube.

Most single-walled nanotubes (SWNTs) have a diameter of close to 1 nanometer, and can be many millions of times longer. The structure of a SWNT can be conceptualized by wrapping a one-atom-thick layer of graphite called graphene into a seamless cylinder. The way the graphene sheet is wrapped is represented by a pair of indices (*n*,*m*). The integers *n* and *m* denote the number of unit vectors along two directions in the honeycomb crystal lattice of graphene. If $m = 0$, the nanotubes are called zigzag nanotubes, and if $n = m$, the nanotubes are called armchair nanotubes. Otherwise, they are called chiral. The diameter of an ideal nanotube can be calculated from its (n,m) indices as follows,

$$d = \frac{a}{\pi}\sqrt{(n^2 + nm + m^2)} = 78.3\sqrt{((n+m)^2 - nm)}\text{pm},$$

where $a = 0.246$ nm.

SWNTs are an important variety of carbon nanotube because most of their properties change significantly with the (*n*,*m*) values, and this dependence is non-monotonic. In par-ticular, their band gap can vary from zero to about 2 eV and their electrical conductivity can show metallic or semiconducting behavior. Single-walled nanotubes are likely candidates for miniatur-izing electronics. The most basic building block of these systems is the electric wire, and SWNTs with diameters of an order of a nanometer can be excellent conductors. One useful application of SWNTs is in the development of the first intermolecular field-effect transistors (FET). The first intermolecular logic gate using SWCNT FETs was made in 2001. A logic gate requires both a p-FET and an n-FET. Because SWNTs are p-FETs when exposed to oxygen and n-FETs otherwise, it is possible to expose half of an SWNT to oxygen and protect the other half from it. The resulting SWNT acts as a *not* logic gate with both p and n-type FETs in the same molecule.

Prices for single-walled nanotubes declined from around $1500 per gram as of 2000 to retail prices of around $50 per gram of as-produced 40–60% by weight SWNTs as of March 2010. As of 2016 the retail price of as-produced 75% by weight SWNTs were $2 per gram, cheap enough for widespread use. SWNTs are forecast to make a large impact in electronics applications by 2020 according to the *The Global Market for Carbon Nanotubes* report.

Multi-walled

Multi-walled nanotubes (MWNTs) consist of multiple rolled layers (concentric tubes) of graphene. There are two models that can be used to describe the structures of multi-walled nanotubes. In the *Russian Doll* model, sheets of graphite are arranged in concentric cylinders, e.g., a (0,8) single-walled nanotube (SWNT) within a larger (0,17) single-walled nanotube. In the *Parchment* model, a single sheet of graphite is rolled in around itself, resembling a scroll of parchment or a rolled newspaper. The interlayer distance in multi-walled nanotubes is close to the distance between graphene layers in graphite, approximately 3.4 Å. The Russian Doll structure is observed more commonly. Its individual shells can be described as SWNTs, which can be metallic or semiconducting. Because of statistical probability and restrictions on the relative diameters of the individual tubes, one of the shells, and thus the whole MWNT, is usually a zero-gap metal.

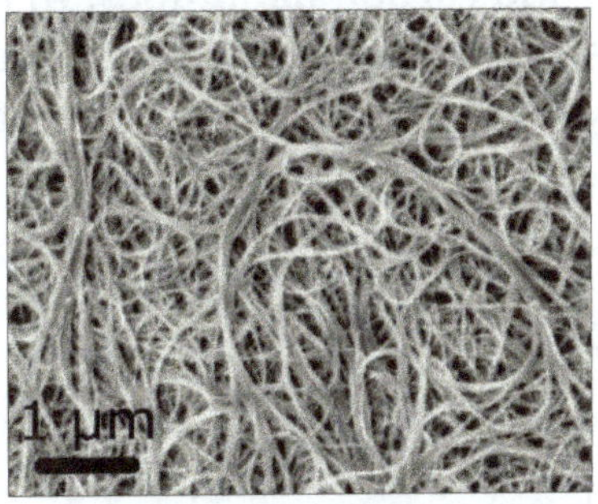

A scanning electron microscopy image of carbon nanotubes bundles.

Double-walled carbon nanotubes (DWNTs) form a special class of nanotubes because their morphology and properties are similar to those of SWNTs but they are more resistant to chemicals. This is especially important when it is necessary to graft chemical functions to the surface of the nanotubes (functionalization) to add properties to the CNT. Covalent functionalization of SWNTs will break some C=C double bonds, leaving "holes" in the structure on the nanotube, and thus modifying both its mechanical and electrical properties. In the case of DWNTs, only the outer wall is modified. DWNT synthesis on the gram-scale was first proposed in 2003 by the CCVD technique, from the selective reduction of oxide solutions in methane and hydrogen.

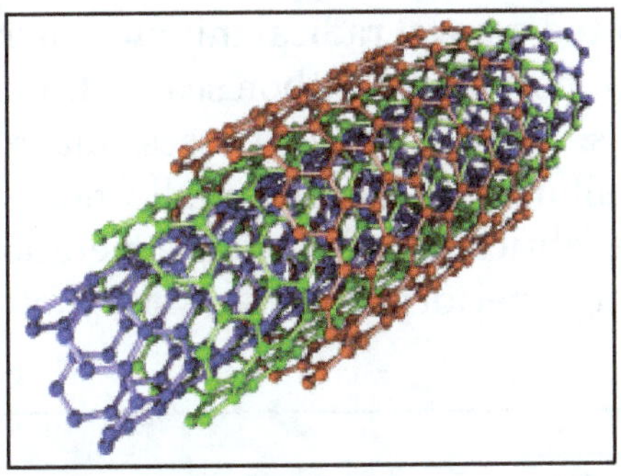

Triple-walled armchair carbon nanotube.

The telescopic motion ability of inner shells and their unique mechanical properties will permit the use of multi-walled nanotubes as main movable arms in coming nanomechanical devices. Retraction force that occurs to telescopic motion caused by the Lennard-Jones interaction between shells and its value is about 1.5 nN.

Torus

In theory, a nanotorus is a carbon nanotube bent into a torus (doughnut shape). Nanotori are predicted to have many unique properties, such as magnetic moments 1000 times larger than previously expected for certain specific radii. Properties such as magnetic moment, thermal stability, etc. vary widely depending on radius of the torus and radius of the tube.

Nanobud

Carbon nanobuds are a newly created material combining two previously discovered allotropes of carbon: carbon nanotubes and fullerenes. In this new material, fullerene-like "buds" are covalently bonded to the outer sidewalls of the underlying carbon nanotube. This hybrid material has useful properties of both fullerenes and carbon nanotubes. In particular, they have been found to be exceptionally good field emitters. In composite materials, the attached fullerene molecules may function as molecular anchors preventing slipping of the nanotubes, thus improving the composite's mechanical properties.

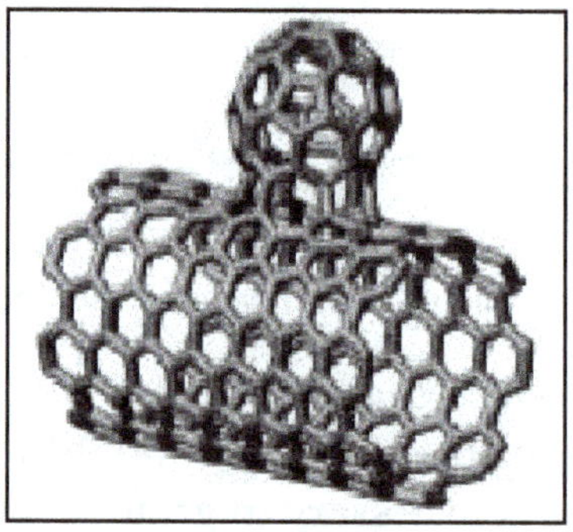

A stable nanobud structure.

Three-dimensional Carbon Nanotube Architectures

Recently, several studies have highlighted the prospect of using carbon nanotubes as building blocks to fabricate three-dimensional macroscopic (>100 nm in all three dimensions) all-carbon devices. Lalwani et al. have reported a novel radical initiated thermal crosslinking method to fabricate macroscopic, free-standing, porous, all-carbon scaffolds using single- and multi-walled carbon nanotubes as building blocks. These scaffolds possess macro-, micro-, and nano- structured pores and the porosity can be tailored for specific applications. These 3D all-carbon scaffolds/architectures may be used for the fabrication of the next generation of energy storage, supercapacitors, field emission transistors, high-performance catalysis, photovoltaics, and biomedical devices and implants.

3D carbon scaffolds.

Graphenated Carbon Nanotubes (g-CNTs)

SEM series of graphenated CNTs with varying foliate density.

Graphenated CNTs are a relatively new hybrid that combines graphitic foliates grown along the sidewalls of multiwalled or bamboo style CNTs. Yu *et al.* reported on "chemically bonded graphene leaves" growing along the sidewalls of CNTs. Stoner *et al.* described these structures as "graphenated CNTs" and reported in their use for enhanced supercapacitor performance. Hsu *et al.* further reported on similar structures formed on carbon fiber paper, also for use in supercapacitor applications. Pham *et al.* also reported a similar structure, namely "graphene-carbon nanotube hybrids", grown directly onto carbon fiber paper to form an integrated, binder free, high surface area conductive catalyst support for Proton Exchange Membrane Fuel Cells electrode applications with enhanced performance and durability. The foliate density can vary as a function of deposition conditions (e.g. temperature and time) with their structure ranging from few layers of graphene (< 10) to thicker, more graphite-like.

The fundamental advantage of an integrated graphene-CNT structure is the high surface area three-dimensional framework of the CNTs coupled with the high edge density of graphene. Graphene edges provide significantly higher charge density and reactivity than the basal plane, but they are difficult to arrange in a three-dimensional, high volume-density geometry. CNTs are readily aligned in a high density geometry (i.e., a vertically aligned forest) but lack high charge density surfaces—the sidewalls of the CNTs are similar to the basal plane of graphene and exhibit low charge density except where edge defects exist. Depositing a high density of graphene foliates along the length of aligned CNTs can significantly increase the total charge capacity per unit of nominal area as compared to other carbon nanostructures.

Nitrogen-doped Carbon Nanotubes

Nitrogen doped carbon nanotubes (N-CNTs) can be produced through five main methods, chemical vapor deposition, high-temperature and high-pressure reactions, gas-solid reaction of amorphous carbon with NH_3 at high temperature, solid reaction, and solvothermal synthesis.

N-CNTs can also be prepared by a CVD method of pyrolyzing melamine under Ar at elevated temperatures of 800–980 °C. However synthesis by CVD of melamine results in the formation of bamboo-structured CNTs. XPS spectra of grown N-CNTs reveal nitrogen in five main components, pyridinic nitrogen, pyrrolic nitrogen, quaternary nitrogen, and nitrogen oxides. Furthermore, synthesis temperature affects the type of nitrogen configuration.

Nitrogen doping plays a pivotal role in lithium storage, as it creates defects in the CNT walls allowing for Li ions to diffuse into interwall space. It also increases capacity by providing more favorable bind of N-doped sites. N-CNTs are also much more reactive to metal oxide nanoparticle deposition which can further enhance storage capacity, especially in anode materials for Li-ion batteries. However boron-doped nanotubes have been shown to make batteries with triple capacity.

Peapod

A carbon peapod is a novel hybrid carbon material which traps fullerene inside a carbon nanotube. It can possess interesting magnetic properties with heating and irradiation. It can also be applied as an oscillator during theoretical investigations and predictions.

Cup-stacked Carbon Nanotubes

Cup-stacked carbon nanotubes (CSCNTs) differ from other quasi-1D carbon structures, which normally behave as quasi-metallic conductors of electrons. CSCNTs exhibit semiconducting behaviors due to the stacking microstructure of graphene layers.

Extreme Carbon Nanotubes

The observation of the *longest* carbon nanotubes grown so far are over 1/2 m (550 mm long) was reported in 2013. These nanotubes were grown on Si substrates using an improved chemical vapor deposition (CVD) method and represent electrically uniform arrays of single-walled carbon nanotubes.

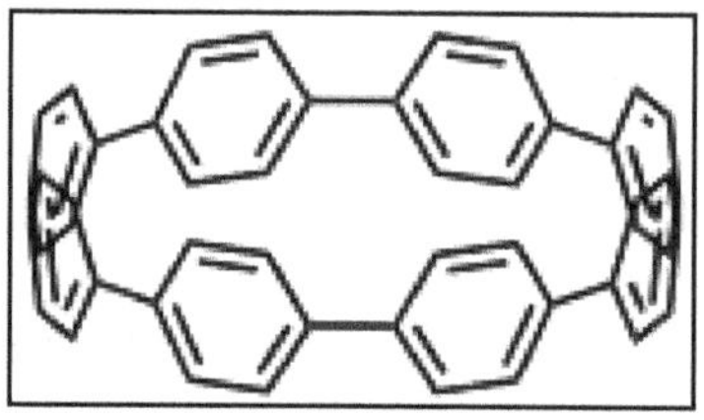

Cycloparaphenylene.

The *shortest* carbon nanotube is the organic compound cycloparaphenylene, which was synthesized in early 2009.

The *thinnest* carbon nanotube is the armchair (2,2) CNT with a diameter of 0.3 nm. This nanotube was grown inside a multi-walled carbon nanotube. Assigning of carbon nanotube type was done by a combination of high-resolution transmission electron microscopy (HRTEM), Raman spectroscopy and density functional theory (DFT) calculations.

The *thinnest freestanding* single-walled carbon nanotube is about 0.43 nm in diameter. Researchers suggested that it can be either (5,1) or (4,2) SWCNT, but the exact type of carbon nanotube remains questionable. (3,3), (4,3) and (5,1) carbon nanotubes (all about 0.4 nm in diameter) were unambiguously identified using aberration-corrected high-resolution transmission electron microscopy inside double-walled CNTs.

The *highest density* of CNTs was achieved in 2013, grown on a conductive titanium-coated copper surface that was coated with co-catalysts cobalt and molybdenum at lower than typical temperatures of 450 °C. The tubes averaged a height of 380 nm and a mass density of 1.6 g cm^{-3}. The material showed ohmic conductivity (lowest resistance 22 kΩ).

Properties

Strength

Carbon nanotubes are the strongest and stiffest materials yet discovered in terms of tensile strength and elastic modulus respectively. This strength results from the covalent sp^2 bonds formed between the individual carbon atoms. In 2000, a multi-walled carbon nanotube was tested to have a tensile strength of 63 gigapascals (9,100,000 psi). (For illustration, this translates into the ability to endure tension of a weight equivalent to 6,422 kilograms-force (62,980 N; 14,160 lbf) on a cable with cross-section of 1 square millimetre (0.0016 sq in).)

Further studies, such as one conducted in 2008, revealed that individual CNT shells have strengths of up to ~100 gigapascals (15,000,000 psi), which is in agreement with quantum/atomistic models. Since carbon nanotubes have a low density for a solid of 1.3 to 1.4 g/cm^3, its specific strength of up to 48,000 kN·m·kg^{-1} is the best of known materials, compared to high-carbon steel's 154 kN·m·kg^{-1}.

Under excessive tensile strain, the tubes will undergo plastic deformation, which means the deformation is permanent. This deformation begins at strains of approximately 5% and can increase the maximum strain the tubes undergo before fracture by releasing strain energy.

Although the strength of individual CNT shells is extremely high, weak shear interactions between adjacent shells and tubes lead to significant reduction in the effective strength of multi-walled carbon nanotubes and carbon nanotube bundles down to only a few GPa. This limitation has been recently addressed by applying high-energy electron irradiation, which crosslinks inner shells and tubes, and effectively increases the strength of these materials to ~60 GPa for multi-walled carbon nanotubes and ~17 GPa for double-walled carbon nanotube bundles.

CNTs are not nearly as strong under compression. Because of their hollow structure and high aspect ratio, they tend to undergo buckling when placed under compressive, torsional, or bending stress.

Comparison of mechanical properties			
Material	**Young's modulus (TPa)**	**Tensile strength (GPa)**	**Elongation at break (%)**
SWNT^E	~1 (from 1 to 5)	13–53	16
Armchair SWNT^T	0.94	126.2	23.1
Zigzag SWNT^T	0.94	94.5	15.6–17.5
Chiral SWNT	0.92		
MWNT^E	0.2–0.8–0.95	11–63–150	
Stainless steel^E	0.186–0.214	0.38–1.55	15–50
Kevlar–29&149^E	0.06–0.18	3.6–3.8	~2

^{E}Experimental observation; ^{T}Theoretical prediction.

The above discussion referred to axial properties of the nanotube, whereas simple geometrical considerations suggest that carbon nanotubes should be much softer in the radial direction than along the tube axis. Indeed, TEM observation of radial elasticity suggested that even the van der Waals forces can deform two adjacent nanotubes. Nanoindentation experiments, performed by several groups on multiwalled carbon nanotubes and tapping/contact mode atomic force microscope measurements performed on single-walled carbon nanotubes, indicated a Young's modulus of the order of several GPa, confirming that CNTs are indeed rather soft in the radial direction.

Hardness

Standard single-walled carbon nanotubes can withstand a pressure up to 25 GPa without [plastic/permanent] deformation. They then undergo a transformation to superhard phase nanotubes. Maximum pressures measured using current experimental techniques are around 55 GPa. However, these new superhard phase nanotubes collapse at an even higher, albeit unknown, pressure.

The bulk modulus of superhard phase nanotubes is 462 to 546 GPa, even higher than that of diamond (420 GPa for single diamond crystal).

Wettability

The surface wettability of CNT is of importance for its applications in various settings. Although the intrinsic contact angle of graphite is around 90°, the contact angles of most as-synthesized CNT arrays are over 160°, exhibiting a superhydrophobic property. By applying a low voltage as low as 1.3V, the extreme water repellant surface can be switched into superhydrophilic.

Kinetic Properties

Multi-walled nanotubes are multiple concentric nanotubes precisely nested within one another. These exhibit a striking telescoping property whereby an inner nanotube core may slide, almost without friction, within its outer nanotube shell, thus creating an atomically perfect linear or rotational bearing. This is one of the first true examples of molecular nanotechnology, the precise positioning of atoms to create useful machines. Already, this property has been utilized to create the world's smallest rotational motor. Future applications such as a gigahertz mechanical oscillator are also envisioned.

Electrical Properties

Unlike graphene, which is a two-dimensional semimetal, carbon nanotubes are either metallic or semiconducting along the tubular axis. For a given (n,m) nanotube, if $n = m$, the nanotube is metallic; if $n - m$ is a multiple of 3, then the nanotube is semiconducting with a very small band gap, otherwise the nanotube is a moderate semiconductor. Thus all armchair $(n = m)$ nanotubes are metallic, and nanotubes (6,4), (9,1), etc. are semiconducting. Carbon nanotubes are not semimetallic because the degenerate point (that point where the π [bonding] band meets the π^* [anti-bonding] band, at which the energy goes to zero) is slightly shifted away from the K point in the Brillouin zone due to the curvature of the tube surface, casing hybridization between the σ^* and π^* anti-bonding bands, modifying the band dispersion.

The rule regarding metallic versus semiconductor behavior has exceptions, because curvature effects in small diameter tubes can strongly influence electrical properties. Thus, a (5,0) SWCNT that should be semiconducting in fact is metallic according to the calculations. Likewise, zigzag and chiral SWCNTs with small diameters that should be metallic have a finite gap (armchair nanotubes remain metallic). In theory, metallic nanotubes can carry an electric current density of 4×10^9 A/cm^2, which is more than 1,000 times greater than those of metals such as copper, where for copper interconnects current densities are limited by electromigration. Carbon nanotubes are thus being explored as conductivity enhancing components in composite materials and many groups are attempting to commercialize highly conducting electrical wire assembled from individual carbon nanotubes. There are significant challenges to be overcome, however, such as the much more resistive nanotube-to-nanotube junctions and impurities, all of which lower the electrical conductivity of the macroscopic nanotube wires by orders of magnitude, as compared to the conductivity of the individual nanotubes.

Because of its nanoscale cross-section, electrons propagate only along the tube's axis. As a result, carbon nanotubes are frequently referred to as one-dimensional conductors. The maximum electrical conductance of a single-walled carbon nanotube is $2G_0$, where $G_0 = 2e^2/h$ is the conductance of a single ballistic quantum channel.

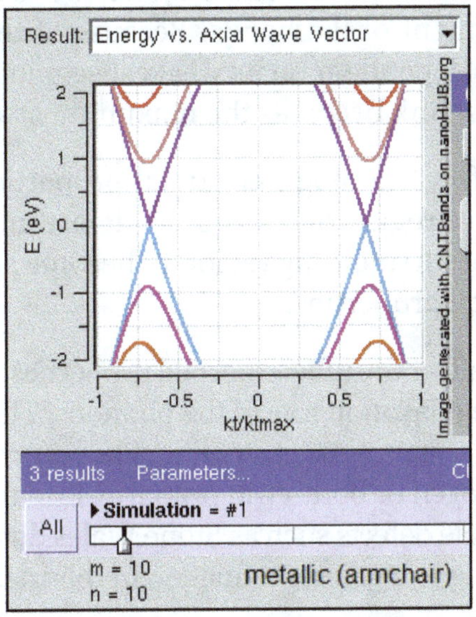

Band structures computed using tight binding approximation for (6,0) CNT (zigzag, metallic), (10,2) CNT (semiconducting) and (10,10) CNT (armchair, metallic).

Due to the role of the π-electron system in determining the electronic properties of graphene, doping in carbon nanotubes differs from that of bulk crystalline semiconductors from the same group of the periodic table (e.g. silicon). Graphitic substitution of carbon atoms in the nanotube wall by boron or nitrogen dopants leads to p-type and n-type behavior, respectively, as would be expected in silicon. However, some non-substitutional (intercalated or adsorbed) dopants introduced into a carbon nanotube, such as alkali metals as well as electron-rich metallocenes, result in n-type conduction because they donate electrons to the π-electron system of the nanotube. By contrast, π-electron acceptors such as $FeCl_3$ or electron-deficient metallocenes function as p-type dopants since they draw π-electrons away from the top of the valence band.

Intrinsic superconductivity has been reported, although other experiments found no evidence of this, leaving the claim a subject of debate.

Optical Properties

Thermal Properties

All nanotubes are expected to be very good thermal conductors along the tube, exhibiting a property known as "ballistic conduction", but good insulators lateral to the tube axis. Measurements show that a SWNT has a room-temperature thermal conductivity along its axis of about 3500 $W{\cdot}m^{-1}{\cdot}K^{-1}$; compare this to copper, a metal well known for its good thermal conductivity, which transmits 385 $W{\cdot}m^{-1}{\cdot}K^{-1}$. A SWNT has a room-temperature thermal conductivity across its axis (in the radial direction) of about 1.52 $W{\cdot}m^{-1}{\cdot}K^{-1}$, which is about as thermally conductive as soil. The temperature stability of carbon nanotubes is estimated to be up to 2800 °C in vacuum and about 750 °C in air.

Defects

As with any material, the existence of a crystallographic defect affects the material properties. Defects can occur in the form of atomic vacancies. High levels of such defects can lower the tensile strength by up to 85%. An important example is the Stone Wales defect, which creates a pentagon and heptagon pair by rearrangement of the bonds. Because of the very small structure of CNTs, the tensile strength of the tube is dependent on its weakest segment in a similar manner to a chain, where the strength of the weakest link becomes the maximum strength of the chain.

Crystallographic defects also affect the tube's electrical properties. A common result is lowered conductivity through the defective region of the tube. A defect in armchair-type tubes (which can conduct electricity) can cause the surrounding region to become semiconducting, and single monatomic vacancies induce magnetic properties.

Crystallographic defects strongly affect the tube's thermal properties. Such defects lead to phonon scattering, which in turn increases the relaxation rate of the phonons. This reduces the mean free path and reduces the thermal conductivity of nanotube structures. Phonon transport simulations indicate that substitutional defects such as nitrogen or boron will primarily lead to scattering of high-frequency optical phonons. However, larger-scale defects such as Stone Wales defects cause phonon scattering over a wide range of frequencies, leading to a greater reduction in thermal conductivity.

Safety and Health

The toxicity of carbon nanotubes has been an important question in nanotechnology. As of 2007, such research had just begun. The data is still fragmentary and subject to criticism. Preliminary results highlight the difficulties in evaluating the toxicity of this heterogeneous material. Parameters such as structure, size distribution, surface area, surface chemistry, surface charge, and agglomeration state as well as purity of the samples, have considerable impact on the reactivity of carbon nanotubes. However, available data clearly show that, under some conditions, nanotubes can cross membrane barriers, which suggests that, if raw materials reach the organs, they can induce harmful effects such as inflammatory and fibrotic reactions.

Synthesis

Techniques have been developed to produce nanotubes in sizable quantities, including arc discharge, laser ablation, high-pressure carbon monoxide disproportionation, and chemical vapor deposition (CVD). Most of these processes take place in a vacuum or with process gases. CVD growth of CNTs can occur in vacuum or at atmospheric pressure. Large quantities of nanotubes can be synthesized by these methods; advances in catalysis and continuous growth are making CNTs more commercially viable.

Chemical Modification

Carbon nanotubes can be functionalized to attain desired properties that can be used in a wide variety of applications. The two main methods of carbon nanotube functionalization are covalent and non-covalent modifications. Because of their hydrophobic nature, carbon nanotubes tend to agglomerate hindering their dispersion is solvents or viscous polymer melts. The resulting nanotube bundles or aggregates reduce the mechanical performance of the final composite. The surface of the carbon nanotubes can be modified to reduce the hydrophobicity and improve interfacial adhesion to a bulk polymer through chemical attachment.

Applications

Current

Current use and application of nanotubes has mostly been limited to the use of bulk nanotubes, which is a mass of rather unorganized fragments of nanotubes. Bulk nanotube materials may never achieve a tensile strength similar to that of individual tubes, but such composites may, nevertheless, yield strengths sufficient for many applications. Bulk carbon nanotubes have already been used as composite fibers in polymers to improve the mechanical, thermal and electrical properties of the bulk product.

- Easton-Bell Sports, Inc. have been in partnership with Zyvex Performance Materials, using CNT technology in a number of their bicycle components–including flat and riser handlebars, cranks, forks, seatposts, stems and aero bars.
- Zyvex Technologies has also built a 54' maritime vessel, the Piranha Unmanned Surface Vessel, as a technology demonstrator for what is possible using CNT technology. CNTs help improve the structural performance of the vessel, resulting in a lightweight 8,000 lb boat that can carry a payload of 15,000 lb over a range of 2,500 miles.
- Amroy Europe Oy manufactures Hybtonite carbon nanoepoxy resins where carbon nanotubes have been chemically activated to bond to epoxy, resulting in a composite material that is 20% to 30% stronger than other composite materials. It has been used for wind turbines, marine paints and variety of sports gear such as skis, ice hockey sticks, baseball bats, hunting arrows, and surfboards.

Other current applications include:

- tips for atomic force microscope probes.
- in tissue engineering, carbon nanotubes can act as scaffolding for bone growth.

There is also ongoing research in using carbon nanotubes as a scaffold for diverse microfabrication techniques.

Potential

The strength and flexibility of carbon nanotubes makes them of potential use in controlling other nanoscale structures, which suggests they will have an important role in nanotechnology engineering. The highest tensile strength of an individual multi-walled carbon nanotube has been tested to be 63 GPa. Carbon nanotubes were found in Damascus steel from the 17th century, possibly helping to account for the legendary strength of the swords made of it. Recently, several studies have highlighted the prospect of using carbon nanotubes as building blocks to fabricate three-dimensional macroscopic (>1mm in all three dimensions) all-carbon devices. Lalwani et al. have reported a novel radical initiated thermal crosslinking method to fabricated macroscopic, free-standing, porous, all-carbon scaffolds using single- and multi-walled carbon nanotubes as building blocks. These scaffolds possess macro-, micro-, and nano- structured pores and the porosity can be tailored for specific applications. These 3D all-carbon scaffolds/architectures maybe used for the fabrication of the next generation of energy storage, supercapacitors, field emission transistors, high-performance catalysis, photovoltaics, and biomedical devices and implants.

Discovery

The true identity of the discoverers of carbon nanotubes is a subject of some controversy. For years, scientists assumed that Sumio Iijima of NEC had discovered carbon nanotubes in 1991. He published a paper describing his discovery which initiated a flurry of excitement and could be credited by inspiring the many scientists now studying applications of carbon nanotubes. Though Iijima has been given much of the credit for discovering carbon nanotubes, it turns out that the timeline of carbon nanotubes goes back much further than 1991. In 1952 L. V. Radushkevich and V. M. Lukyanovich published clear images of 50 nanometer diameter tubes made of carbon in the Soviet *Journal of Physical Chemistry*. This discovery was largely unnoticed, as the article was published in Russian, and Western scientists' access to Soviet press was limited during the Cold War. Before they came to be known as carbon nanotubes, in 1976, Morinobu Endo of CNRS observed hollow tubes of rolled up graphite sheets synthesised by a chemical vapour-growth technique. The first specimens observed would later come to be known as single-walled carbon nanotubes (SWNTs). The three scientists have been the first ones to show images of a nanotube with a solitary graphene wall.

Endo, in his early review of vapor-phase-grown carbon fibers (VPCF), also reminded us that he had observed a hollow tube, linearly extended with parallel carbon layer faces near the fiber core. This appears to be the observation of multi-walled carbon nanotubes at the center of the fiber. The mass-produced MWCNTs today are strongly related to the VPGCF developed by Endo. In fact, they call it the "Endo-process", out of respect for his early work and patents.

In 1979, John Abrahamson presented evidence of carbon nanotubes at the 14th Biennial Conference of Carbon at Pennsylvania State University. The conference paper described carbon nanotubes as carbon fibers that were produced on carbon anodes during arc discharge. A characterization of these fibers was given as well as hypotheses for their growth in a nitrogen atmosphere at low pressures.

In 1981, a group of Soviet scientists published the results of chemical and structural characterization of carbon nanoparticles produced by a thermocatalytical disproportionation of carbon monoxide. Using TEM images and XRD patterns, the authors suggested that their "carbon multi-layer tubular crystals" were formed by rolling graphene layers into cylinders. They speculated that by rolling graphene layers into a cylinder, many different arrangements of graphene hexagonal nets are possible. They suggested two possibilities of such arrangements: circular arrangement (armchair nanotube) and a spiral, helical arrangement (chiral tube).

In 1987, Howard G. Tennent of Hyperion Catalysis was issued a U.S. patent for the production of "cylindrical discrete carbon fibrils" with a "constant diameter between about 3.5 and about 70 nanometers..., length 10^2 times the diameter, and an outer region of multiple essentially continuous layers of ordered carbon atoms and a distinct inner core...."

Iijima's discovery of multi-walled carbon nanotubes in the insoluble material of arc-burned graphite rods in 1991 and Mintmire, Dunlap, and White's independent prediction that if single-walled carbon nanotubes could be made, then they would exhibit remarkable conducting properties helped create the initial buzz that is now associated with carbon nanotubes. Nanotube research accelerated greatly following the independent discoveries by Bethune at IBM and Iijima at NEC of *single-walled* carbon nanotubes and methods to specifically produce them by adding transition-metal catalysts to the carbon in an arc discharge. The arc discharge technique was well-known to produce the famed Buckminster fullerene on a preparative scale, and these results appeared to extend the run of accidental discoveries relating to fullerenes. The discovery of nanotubes remains a contentious issue. Many believe that Iijima's report in 1991 is of particular importance because it brought carbon nanotubes into the awareness of the scientific community as a whole.

Silver Nanoparticle

Silver nanoparticles are nanoparticles of silver of between 1 nm and 100 nm in size. While frequently described as being 'silver' some are composed of a large percentage of silver oxide due to their large ratio of surface-to-bulk silver atoms. Numerous shapes of nanoparticles can be constructed depending on the application at hand. Commonly used are spherical silver nanoparticles but diamond, octagonal and thin sheets are also popular.

Their extremely large surface area permits the coordination of a vast number of ligands. The properties of silver nanoparticles applicable to human treatments are under investigation in laboratory and animal studies, assessing potential efficacy, toxicity, and costs.

Synthesis

Methods

Silver nanoparticles can be synthesized by several methods. They can be divided into three broad categories: wet chemistry, ion implantation, or biogenic synthesis.

Wet Chemistry

Several wet chemical methods have been developed to form silver nanoparticles. Each of these methods use the same basic mechanism: a silver ion complex is reduced to colloidal silver in the presence of a reducing agent, which oxidizes and gives back the missing electron, so that the silver ion is reduced back to its metallic state. The newly formed silver nanoparticles can then also be stabilized with a stabilizing/capping agent, which stabilizes the energy of the surface of the newly formed nanoparticle to prevent them from aggregate to larger particles. These methods include the use of reducing sugars, citrate reduction, reduction via sodium borohydride, the silver mirror reaction, the polyol process, seed-mediated growth, and light-mediated growth. Each of these methods, or combination of methods, will offer differing degrees of control over the size distribution as well as distributions of geometric arrangements of the nanoparticle. A new, very promising wet-chemical technique was found by Elsupikhe et al. (2015). They have developed a green ultrasonically-assisted synthesis. Under ultrasound treatment, silver nanoparticles (AgNP) are synthesized with κ-carrageenan as a natural stabilizer. The reaction is performed at ambient temperature and produces silver nanoparticles with fcc crystal structure without impurities. The concentration of κ-carrageenan is used to influence particle size distribution of the AgNPs.

Reduction with Reducing Sugars

Reducing sugars includes glucose, fructose, maltose, maltodextrin, etc. but not sucrose, and is the simpliest method to reduce silver ions back to silver nanoparticles. And if the pH is over 7 the glucose molucule in the reducing sugar will "open up" its ring structure and attach to the surface of the nanoparticles, and thereby act as a weak stabilizing agent as well.

Citrate Reduction

An early, and very common, method for synthesizing silver nanoparticles is citrate reduction. Citrate reduction involves the reduction of silver nitrate, $AgNO_3$, to colloidal silver using trisodium citrate, $Na_3C_6H_5O_7$ at elevated temperature (~100 °C). In this synthesis, the citrate ion acts as both the reducing and capping agent. This method is useful due to its relative ease and short reaction time. However, the silver particles formed may exhibit broad size distributions and/or form several different particle geometries simultaneously.

Reduction Via Sodium Borohydride

A similar method is the reduction of silver nitrate using sodium borohydride, NaBH4. Sodium borohydride is a stronger reducing agent than citrate, but it is not thought to participate in surface stabilization. Thus, a separate capping molecule must be used in addition to the sodium borohydride. Examples of stabilizers used include citrate, poly(vinyl pyrrolidone) (PVP), bovine serum albumin, and cetyltrimethylammonium bromide (CTAB).

Polyol Process

The polyol process is a particularly useful method because it yields a high degree of control over both the size and geometry of the resulting nanoparticles. In general, the polyol synthesis begins with the heating of a polyol compound such as ethylene glycol, 1,5-pentanediol, or 1,2-propylene glycol7. An Ag+ species and a capping agent are added (although the polyol itself is also often the capping agent). The Ag+ species is then reduced by the polyol to colloidal nanoparticles. The polyol process is highly sensitive to reaction conditions such as temperature, chemical environment, and concentration of substrates.(add refs again here) Therefore, by changing these variables, various sizes and geometries can be selected for such as quasi-spheres, pyramids, spheres, and wires. Further study has examined the mechanism for this process as well as resulting geometries under various reaction conditions in greater detail.

Seed-mediated Growth

Seed-mediated growth is the process in which seed nanoparticles, formed from a reduction of Ag+ to colloidal silver by any method, are placed in a growth solution. The growth solution contains an additional silver species that deposits more silver atoms onto the original seeds. Several experimental factors are manipulated to achieve size and geometry control in this method such as concentration of the substrates and the capping agent used. Parameters controlling the size and geometry of the particles is described in greater detail in various studies.

Light-mediated Growth

Light-mediated syntheses have also been explored where light can promote formation of various silver nanoparticle morphologies.

Silver Mirror Reaction

The silver mirror reaction involves the conversion of silver nitrate to Ag(NH3)OH. Ag(NH3)OH is subsequently reduced into colloidal silver using an aldehyde containing molecule such as a sugar. The silver mirror reaction is as follows:

$$2(Ag(NH_3)_2)^+ + RCHO + 2OH^- RCOOH + 2Ag + 4NH_3$$

The size and shape of the nanoparticles produced are difficult to control and often have wide distributions. However, this method is often used to apply thin coatings of silver particles onto surfaces and further study into producing more uniformly sized nanoparticles is being done.

Ion Implantation

Ion implantation has been used to create silver nanoparticles embedded in glass, polyurethane, silicone, polyethylene, and poly(methyl methacrylate). Particles are embedded in the substrate by means of bombardment at high accelerating voltages. At a fixed current density of the ion beam up to a certain value, the size of the embedded silver nanoparticles has been found to be monodisperse within the population, after which only an increase in the ion concentration is observed. A further increase in the ion beam dose has been found to reduce both the nanoparticle size and density in the target substrate, whereas an ion beam operating at a high accelerating voltage with a gradually increasing current density has been found to result in a gradual increase in the nanoparticle size. There are a few competing mechanisms which may result in the decrease in nanoparticle size; destruction of NPs upon collision, sputtering of the sample surface, particle fusion upon heating and dissociation.

The formation of embedded nanoparticles is complex, and all of the controlling parameters and factors have not yet been investigated. Computer simulation is still difficult as it involves processes of diffusion and clustering, however it can be broken down into a few different sub-processes such as implantation, diffusion, and growth. Upon implantation, silver ions will reach different depths within the substrate which approaches a Gaussian distribution with the mean centered at X depth. High temperature conditions during the initial stages of implantation will increase the impurity diffusion in the substrate and as a result limit the impinging ion saturation, which is required for nanoparticle nucleation. Both the implant temperature and ion beam current density are crucial to control in order to obtain a monodisperse nanoparticle size and depth distribution. A low current density may be used to counter the thermal agitation from the ion beam and a buildup of surface charge. After implantation on the surface, the beam currents may be raised as the surface conductivity will increase. The rate at which impurities diffuse drops quickly after the formation of the nanoparticles, which act as a mobile ion trap. This suggests that the beginning of the implantation process is critical for control of the spacing and depth of the resulting nanoparticles, as well as control of the substrate temperature and ion beam density. The presence and nature of these particles can be analyzed using numerous spectroscopy and microscopy instruments. Nanoparticles synthesized in the substrate exhibit surface plasmon resonances as evidenced by characteristic absorption bands; these features undergo spectral shifts depending on the nanoparticle size and surface asperities, however the optical properties also strongly depend on the substrate material of the composite.

Biogenic Synthesis

The biological synthesis of nanoparticles has provided a means for improved techniques compared to the traditional methods that call for the use of harmful reducing agents like sodium borohydride. Many of these methods could improve their environmental footprint by replacing these relatively strong reducing agents. The problems with the chemical production of silver nanoparticles is usually involves high cost and the longevity of the particles is short lived due to aggregation. The harshness of standard chemical methods has sparked the use of using biological organisms to reduce silver ions in solution into colloidal nanoparticles.

In addition, precise control over shape and size is vital during nanoparticle synthesis since the NPs therapeutic properties are intimately dependent on such factors. Hence, the primary focus of research in biogenic synthesis is in developing methods that consistently reproduce NPs with precise properties.

Use of Fungi and Bacteria

A general representation of the synthesis and applications of biogenically synthesized silver nanoparticles using plant extract.

Bacterial and fungal synthesis of nanoparticles is practical because bacteria and fungi are easy to handle and can be modified genetically with ease. This provides a means to develop biomolecules that can synthesize AgNPs of varying shapes and sizes in high yield, which is at the forefront of current challenges in nanoparticle synthesis. Fungal strains such as Verticillium and bacterial strains such as K. pneumoniae can be used in the synthesis of silver nanoparticles. When the fungus/bacteria is added to solution, protein biomass is released into the solution. Electron donating residues such as tryptophan and tyrosine reduce silver ions in solution contributed by silver nitrate. These methods have been found to effectively create stable monodisperse nanoparticles without the use of harmful reducing agents.

A method has been found of reducing silver ions by the introduction of the fungus *Fusarium oxysporum*. The nanoparticles formed in this method have a size range between 5 and 15 nm and consist of silver hydrosol. The reduction of the silver nanoparticles is thought to come from an enzymatic process and silver nanoparticles produced are extremely stable due to interactions with proteins that are excreted by the fungi.

Bacterium found in silver mines, Pseudomonas stutzeri AG259, were able to construct silver particles in the shapes of triangles and hexagons. The size of these nanoparticles had a large range in size and some of them reached sizes larger than the usual nanoscale with a size of 200 nm. The silver nanoparticles were found in the organic matrix of the bacteria.

Lactic acid producing bacteria have been used to produce silver nanoparticles. The bacteria *Lactobacillus* spp., *Pediococcus pentosaceus, Enteroccus faeciumI*, and *Lactococcus garvieae* have been found to be able to reduce silver ions into silver nanoparticles. The production of the nanoparticles takes place in the cell from the interactions between the silver ions and the organic compounds of the cell. It was found that the bacterium *Lactobacillus fermentum* created the smallest silver nanoparticles with an average size of 11.2 nm. It was also found that this bacterium produced the nanoparticles with the smallest size distribution and the nanoparticles were found mostly on the outside of the cells. It was also found that there was an increase in the pH increased the rate of which the nanoparticles were produced and the amount of particles produced.

Use of Plants

The reduction of silver ions into silver nanoparticles has also been achieved using geranium leaves. It has been found that adding geranium leaf extract to silver nitrate solutions causes there silver ions to be quickly reduced and that the nanoparticles produced are particularly stable. The silver nanoparticles produced in solution had a size range between 16 and 40 nm.

In another study different plant leaf extracts were used to reduce silver ions. It was found that out of Camellia sinensis (green tea), pine, persimmon, ginko, magnolia, and platanus that the magnolia leaf extract was the best at creating silver nanoparticles. This method created particles with a disperse size range of 15 to 500 nm, but it was also found that the particle size could be controlled by varying the reaction temperature. The speed at which the ions were reduced by the magnolia leaf extract was comparable to those of using chemicals to reduce.

The use of plants, microbes, and fungi in the production of silver nanoparticles is leading the way to more environmentally sound production of silver nanoparticles.

Products and Functionalization

Synthetic protocols for silver nanoparticle production can be modified to produce silver nanoparticles with non-spherical geometries and also to functionalize nanoparticles with different materials, such as silica. Creating silver nanoparticles of different shapes and surface coatings allows for greater control over their size-specific properties.

Anisotropic Structures

Silver nanoparticles can be synthesized in a variety of non-spherical (anisotropic) shapes. Because silver, like other noble metals, exhibits a size and shape dependent optical effect known as localized surface plasmon resonance (LSPR) at the nanoscale, the ability to synthesize Ag nanoparticles in different shapes vastly increases the ability to tune their optical behavior. For example, the wavelength at which LSPR occurs for a nanoparticle of one morphology (e.g. a sphere) will be different if that sphere is changed into a different shape. This shape dependence allows a silver nanoparticle to experience optical enhancement at a range of different wavelengths, even by keeping the size relatively constant, just by changing its shape. The applications of this shape-exploited expansion of optical behavior range from developing more sensitive biosensors to increasing the longevity of textiles.

Triangular Nanoprisms

Triangular shaped nanoparticles are a canonical type of anisotropic morphology studied for both gold and silver.

Though many different techniques for silver nanoprism synthesis exist, several methods employ a seed-mediated approach, which involves first synthesizing small (3-5 nm diameter) silver nanoparticles that offer a template for shape-directed growth into triangular nanostructures.

The silver seeds are synthesized by mixing silver nitrate and sodium citrate in aqueous solution and then rapidly adding sodium borohydride. Additional silver nitrate is added to the seed solution at low temperature, and the prisms are grown by slowly reducing the excess silver nitrate using ascorbic acid.

With the seed-mediated approach to silver nanoprism synthesis, selectivity of one shape over another can in part be controlled by the capping ligand. Using essentially the same procedure above but changing citrate to poly (vinyl pyrrolidone) (PVP) yields cube and rod-shaped nanostructures instead of triangular nanoprisms.

In addition to the seed mediated technique, silver nanoprisms can also be synthesized using a photo-mediated approach, in which preexisting spherical silver nanoparticles are transformed into triangular nanoprisms simply by exposing the reaction mixture to high intensities of light.

Nanocubes

Silver nanocubes can be synthesized using ethylene glycol as a reducing agent and PVP as a capping agent, in a polyol synthesis reaction (vide supra). A typical synthesis using these reagents involves adding fresh silver nitrate and PVP to a solution of ethylene glycol heated at 140 °C.

This procedure can actually be modified to produce another anisotropic silver nanostructure, nanowires, by just allowing the silver nitrate solution to age before using it in the synthesis. By allowing the silver nitrate solution to age, the initial nanostructure formed during the synthesis is slightly different than that obtained with fresh silver nitrate, which influences the growth process, and therefore, the morphology of the final product.

Coating with Silica

In this method, polyvinylpyrrolidone (PVP) is dissolved in water by sonication and mixed with silver colloid particles. Active stirring ensures the PVP has adsorbed to the nanoparticle surface. Centrifuging separates the PVP coated nanoparticles which are then transferred to a solution of ethanol to be centrifuged further and placed in a solution of ammonia, ethanol and $Si(OEt_4)$ (TES). Stirring for twelve hours results in the silica shell being formed consisting of a surrounding layer of silicon oxide with an ether linkage available to add functionality. Varying the amount of TES allows for different thicknesses of shells formed. This technique is popular due to the ability to add a variety of functionality to the exposed silica surface.

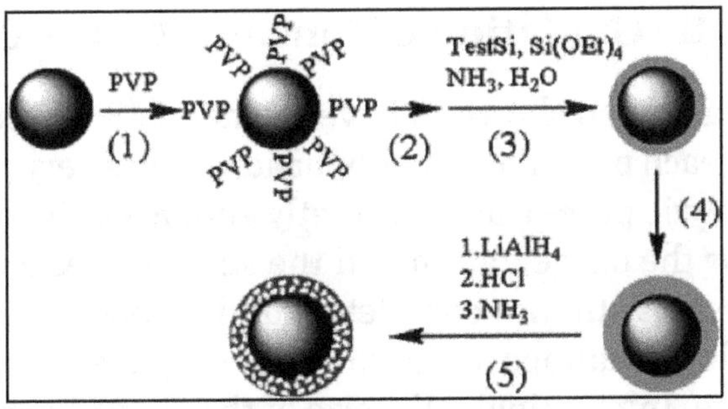

General procedure for coating colloid particles in silica. First PVP is absorbed onto the colloidal surface. These particles are put into a solution of ammonia in ethanol. the particle then begins to grow by addition of Si(OET4).

Use

Catalysis

Using silver nanoparticles for catalysis has been gaining attention in recent years. Although the most common applications are for medicinal or antibacterial purposes, silver nanoparticles have been demonstrated to show catalytic redox properties for dyes, benzene, carbon monoxide, and likely other compounds.

NOTE: This paragraph is a general description of nanoparticle properties for catalysis; it is not exclusive to silver nanoparticles. The size of a nanoparticle greatly determines the properties that it exhibits due to various quantum effects. Additionally, the chemical environment of the nanoparticle plays a large role on the catalytic properties. With this in mind, it is important to note that heterogeneous catalysis takes place by adsorption of the reactant species to the catalytic substrate. When polymers, complex ligands, or surfactants are used to prevent coalescence of the nanoparticles, the catalytic ability is frequently hindered due to reduced adsorption ability. However, these compounds can also be used in such a way that the chemical environment enhances the catalytic ability.

Supported on Silica Spheres – Reduction of Dyes

Silver nanoparticles have been synthesized on a support of inert silica spheres. The support plays virtually no role in the catalytic ability and serves as a method of preventing coalescence of the silver nanoparticles in colloidal solution. Thus, the silver nanoparticles were stabilized and it was possible to demonstrate the ability of them to serve as an electron relay for the reduction of dyes by sodium borohydride. Without the silver nanoparticle catalyst, virtually no reaction occurs between sodium borohydride and the various dyes: methylene blue, eosin, and rose Bengal.

Mesoporous Aerogel – Selective Oxidation of Benzene

Silver nanoparticles supported on aerogel are advantageous due to the higher number of active sites. The highest selectivity for oxidation of benzene to phenol was observed at low weight percent of silver in the aerogel matrix (1% Ag). This better selectivity is believed to be a result of the higher monodispersity within the aerogel matrix of the 1% Ag sample. Each weight percent solution formed different sized particles with a different width of size range.

Silver Alloy – Synergistic Oxidation of Carbon Monoxide

Au-Ag alloy nanoparticles have been shown to have a synergistic effect on the oxidation of carbon monoxide (CO). On its own, each pure-metal nanoparticle shows very poor catalytic activity for CO oxidation; together, the catalytic properties are greatly enhanced. It is proposed that the gold acts as a strong binding agent for the oxygen atom and the silver serves as a strong oxidizing catalyst, although the exact mechanism is still not completely understood. When synthesized in an Au/Ag ratio from 3:1 to 10:1, the alloyed nanoparticles showed complete conversion when 1% CO was fed in air at ambient temperature. Interestingly, the size of the alloyed particles did not play a big role in the catalytic ability. It is well known that gold nanoparticles only show catalytic properties for CO when they are ~3 nm in size, but alloyed particles up to 30 nm demonstrated excellent catalytic activity – catalytic activity better than that of gold nanoparticles on active support such as TiO2, Fe2O3, etc.

Light-enhanced

Plasmonic effects have been studied quite extensively. Until recently, there have not been studies investigating the oxidative catalytic enhancement of a nanostructure via excitation of its surface plasmon resonance. The defining feature for enhancing the oxidative catalytic ability has been identified as the ability to convert a beam of light into the form of energetic electrons that can be transferred to adsorbed molecules. The implication of such a feature is that photochemical reactions can be driven by low-intensity continuous light can be coupled with thermal energy.

The coupling of low-intensity continuous light and thermal energy has been performed with silver nanocubes. The important feature of silver nanostructures that are enabling for photocatalysis is their nature to create resonant surface plasmons from light in the visible range.

The addition of light enhancement enabled the particles to perform to the same degree as particles that were heated up to 40K greater. This is a profound finding when noting that a reduction in temperature of 25K can increase the catalyst lifetime by nearly tenfold, when comparing the photothermal and thermal process.

Biological Research

Researchers have explored the use of silver nanoparticles as carriers for delivering various payloads such as small drug molecules or large biomolecules to specific targets. Once the AgNP has had sufficient time to reach its target, release of the payload could potentially be triggered by an internal or external stimulus. The targeting and accumulation of nanoparticles may provide high payload concentrations at specific target sites and could minimize side effects.

Chemotherapy

The introduction of nanotechnology into medicine is expected to advance diagnostic cancer imaging and the standards for therapeutic drug design. Nanotechnology may uncover insight about the structure, function and organizational level of the biosystem at the nanoscale.

Silver nanoparticles can undergo coating techniques that offer a uniform functionalized surface to which substrates can be added. When the nanoparticle is coated, for example, in silica the surface exists as silicic acid. Substrates can thus be added through stable ether and ester linkages that are not degraded immediately by natural metabolic enzymes. Recent chemotherapeutic applications have designed anti cancer drugs with a photo cleavable linker, such as an ortho-nitrobenzyl bridge, attaching it to the substrate on the nanoparticle surface. The low toxicity nanoparticle complex can remain viable under metabolic attack for the time necessary to be distributed throughout the bodies systems. If a cancerous tumor is being targeted for treatment, ultraviolet light can be introduced over the tumor region. The electromagnetic energy of the light causes the photo responsive linker to break between the drug and the nanoparticle substrate. The drug is now cleaved and released in an unaltered active form to act on the cancerous tumor cells. Advantages anticipated for this method is that the drug is transported without highly toxic compounds, the drug is released without harmful radiation or relying on a specific chemical reaction to occur and the drug can be selectively released at a target tissue.

A second approach is to attach a chemotherapeutic drug directly to the functionalized surface of the silver nanoparticle combined with a nucelophilic species to undergo a displacement reaction. For example, once the nanoparticle drug complex enters or is in the vicinity of the target tissue or cells, a glutathione monoester can be administered to the site. The nucleophilic ester oxygen will attach to the functionalized surface of the nanoparticle through a new ester linkage while the drug is released to its surroundings. The drug is now active and can exert its biological function on the cells immediate to its surroundings limiting non-desirable interactions with other tissues.

Multiple Drug Resistance

A major cause for the ineffectiveness of current chemotherapy treatments is multiple drug resistance which can arise from several mechanisms.

Nanoparticles can provide a means to overcome MDR. In general, when using a targeting agent to deliver nanocarriers to cancer cells, it is imperative that the agent binds with high selectivity to molecules that are uniquely expressed on the cell surface. Hence NPs can be designed with proteins that specifically detect drug resistant cells with overexpressed transporter proteins on their surface. A pitfall of the commonly used nano-drug delivery systems is that free drugs that are released from the nanocarriers into the cytosol get exposed to the MDR transporters once again, and are exported. To solve this, 8 nm nano crystalline silver particles were modified by the addition of trans-activating transcriptional activator (TAT), derived from the HIV-1 virus, which acts as a cell penetrating peptide (CPP). Generally, AgNP effectiveness is limited due to the lack of efficient cellular uptake; however, CPP-modification has become one of the most efficient methods for improving intracellular delivery of nanoparticles. Once ingested, the export of the AgNP is prevented based on a size exclusion. The concept is simple: the nanoparticles are too large to be effluxed by the MDR transporters, because the efflux function is strictly subjected to the size of its substrates, which is generally limited to a range of 300-2000 Da. Thereby the nanoparticulates remain insusceptible to the efflux, providing a means to accumulate in high concentrations.

Antimicrobial

Introduction of silver into bacterial cells induces a high degree of structural and morphological changes, which can lead to cell death. As the silver nano particles come in contact with the bacteria, they adhere to the cell wall and cell membrane. Once bound, some of the silver passes through to the inside, and interacts with phosphate-containing compounds like DNA and RNA, while another portion adheres to the sulphur-containing proteins on the membrane. The silver-sulphur interactions at the membrane cause the cell wall to undergo structural changes, like the formation of pits and pores. Through these pores, cellular components are released into the extracellular fluid, simply due to the osmotic difference. Within the cell, the integration of silver creates a low molecular weight region where the DNA then condenses. Having DNA in a condensed state inhibits the cell's replication proteins contact with the DNA. Thus the introduction of silver nanoparticles inhibits replication and is sufficient to cause the death of the cell. Further increasing their effect, when silver comes in contact with fluids, it tends to ionize which increases the nanoparticles bactericidal activity. This has been correlated to the suppression of enzymes and inhibited expression of proteins that relate to the cell's ability to produce ATP.

Although it varies for every type of cell proposed, as their cell membrane composition varies greatly, It has been seen that in general, silver nano particles with an average size of 10 nm or less show electronic effects that greatly increase their bactericidal activity. This could also be partly due to the fact that as particle size decreases, reactivity increases due to the surface area to volume ratio increasing.

It has been noted that the introduction of silver nano particles has shown to have synergistic activity with common antibiotics already used today, such as; penicillin G, ampicillin, erythromycin, clindamycin, and vancomycin against E. coli and S. aureus. In medical equipment, it has been shown that silver nano particles drastically lower the bacterial count on devices used. However, the problem arises when the procedure is over and a new one must be done. In the process of washing the instruments a large portion of the silver nano particles become less effective due to the loss of silver ions. They are more commonly used in skin grafts for burn victims as the silver nano particles embedded with the graft provide better antimicrobial activity and result in significantly less scarring of the victim. They also show promising application as water treatment method to form clean potable water.

Other

Samsung has created and marketed a material called Silver Nano, that includes silver nanoparticles on the surfaces of household appliances.

Safety

Although silver nanoparticles are widely used in a variety of commercial products, there has only recently been a major effort to study their effects on human health. There have been several studies that describe the *in vitro* toxicity of silver nanoparticles to a variety of different organs, including the lung, liver, skin, brain, and reproductive organs. The mechanism of the toxicity of silver nanoparticles to human cells appears to be derived from oxidative stress and inflammation that is caused by the generation of reactive oxygen species (ROS) stimulated by either the Ag NPs, Ag ions, or both. For example, Park *et al.* showed that exposure of a mouse peritoneal macrophage cell line (RAW267.7) to silver nanoparticles decreased the cell viability in a concentration- and time-dependent manner. They further showed that the intracellular reduced glutathionine (GSH), which is a ROS scavenger, decreased to 81.4% of the control group of silver nanoparticles at 1.6 ppm.

Modes of Toxicity

Since silver nanoparticles undergo dissolution releasing silver ions, which is well-documented to have toxic effects, there have been several studies that have been conducted to determine whether the toxicity of silver nanoparticles is derived from the release of silver ions or from the nanoparticle itself. Several studies suggest that the toxicity of silver nanoparticles is attributed to their release of silver ions in cells as both silver nanoparticles and silver ions have been reported to have similar cytotoxicity. For example, In some cases it is reported that silver nanoparticles facilitate the release of toxic free silver ions in cells via a "Trojan-horse type mechanism," where the particle enters cells and is then ionized within the cell. However, there have been reports that suggest that a combination of silver nanoparticles and ions is responsible for the toxic effect of silver nanoparticles. Navarro *et al.* using cysteine ligands as a tool to measure the concentration of free silver in solution, determined that although initially silver ions were 18 times more likely to inhibit the photosynthesis of an algae, Chlamydomanas reinhardtii, but after 2 hours of incubation it was revealed that the algae containing silver nanoparticles were more toxic than just silver ions alone. Furthermore, there are studies that suggest that silver nanoparticles induce toxicity independent of free silver ions. For example, Asharani *et al.* compared phenotypic defects observed in zebrafish treated with silver nanoparticles and silver ions and determined that the phenotypic defects observed with silver nanoparticle treatment was not observed with silver ion-treated embryos, suggesting that the toxicity of silver nanoparticles are independent of silver ions.

Protein channels and nuclear membrane pores can often be in the size range of 9 nm to 10 nm in diameter. Small silver nanoparticles constructed of this size have the ability to not only pass through the membrane to interact with internal structures but also to be become lodged within the membrane. Silver nanoparticle depositions in the membrane can impact regulation of solutes, exchange of proteins and cell recognition. Exposure to silver nanoparticles has been associated with "inflammatory, oxidative, genotoxic, and cytotoxic consequences"; the silver particulates primarily accumulate in the liver. but have also been shown to be toxic in other organs including the brain. Nano-silver applied to tissue-cultured human cells leads to the formation of free radicals, raising concerns of potential health risks.

- Allergic reaction: There have been several studies conducted that show a precedence for allerginicity of silver nanoparticles.
- Argyria and staining: Ingested silver or silver compounds, including colloidal silver, can cause a condition called argyria, a discoloration of the skin and organs.In 2006, there was a case study of a 17-year-old man, who sustained burns to 30% of his body, and experienced a temporary bluish-grey hue after several days of treatment with Acticoat, a brand of wound dressing containing silver nanoparticles. Argyria is the deposition of silver in deep tissues, a condition that cannot happen on a temporary basis, raising the question of whether the cause of the man's discoloration was argyria or even a result of the silver treatment. Silver dressings are known to cause a "transient discoloration" that dissipates in 2–14 days, but not a permanent discoloration.
- Silzone heart valve: St. Jude Medical released a mechanical heart valve with a silver coated sewing cuff (coated using ion beam-assisted deposition) in 1997. The valve was designed to reduce the instances of endocarditis. The valve was approved for sale in Canada, Europe,

the United States, and most other markets around the world. In a post-commercialization study, researchers showed that the valve prevented tissue ingrowth, created paravalvular leakage, valve loosening, and in the worst cases explantation. After 3 years on the market and 36,000 implants, St. Jude discontinued and voluntarily recalled the valve.

Hybrid Material

Hybrid materials are composites consisting of two constituents at the nanometer or molecular level. Commonly one of these compounds is inorganic and the other one organic in nature. Thus, they differ from traditional composites where the constituents are at the macroscopic (micrometer to millimeter) level. Mixing at the microscopic scale leads to a more homogeneous material that either show characteristics in between the two original phases or even new properties.

Introduction

Hybrid Materials in Nature

Many natural materials consist of inorganic and organic building blocks distributed on the nanoscale. In most cases the inorganic part provides mechanical strength and an overall structure to the natural objects while the organic part delivers bonding between the inorganic building blocks and/or the soft tissue. Typical examples of such materials are bone, or nacre.

Development of Hybrid Materials

The first hybrid materials were the paints made from inorganic and organic components that were used thousands of years ago. Rubber is an example of the use of inorganic materials as fillers for organic polymers. The sol–gel process developed in the 1930s was one of the major driving forces what has become the broad field of inorganic–organic hybrid materials.

Classification

Hybrid materials can be classified based on the possible interactions connecting the inorganic and organic species. *Class I* hybrid materials are those that show weak interactions between the two phases, such as van der Waals, hydrogen bonding or weak electrostatic interactions. *Class II* hybrid materials are those that show strong chemical interactions between the components such as covalent bonds.

Structural properties can also be used to distinguish between various hybrid materials. An organic moiety containing a functional group that allows the attachment to an inorganic network, e.g. a trialkoxysilane group, can act as a *network modifier* because in the final structure the inorganic network is only modified by the organic group. Phenyltrialkoxysilanes are an example for such compounds; they modify the silica network in the sol–gel process via the reaction of the trialkoxysilane group without supplying additional functional groups intended to undergo further chemical reactions to the material formed. If a reactive functional group is incorporated the system is called a *network functionalizer*. The situation is different if two or three of such anchor groups modify an organic segment; this leads to materials in which the inorganic group is afterwards an integral part of the hybrid network. The latter type of system is known as *network builder*.

Blends are formed if no strong chemical interactions exist between the inorganic and organic building blocks. One example for such a material is the combination of inorganic clusters or particles with organic polymers lacking a strong (e.g. covalent) interaction between the components. In this case a material is formed that consists for example of an organic polymer with entrapped discrete inorganic moieties in which, depending on the functionalities of the components, for example weak crosslinking occurs by the entrapped inorganic units through physical interactions or the inorganic components are entrapped in a crosslinked polymer matrix. If an inorganic and an organic network interpenetrate each other without strong chemical interactions, so called interpenetrating networks (IPNs) are formed, which is for example the case if a sol–gel material is formed in presence of an organic polymer or vice versa. Both materials described belong to class I hybrids. Class II hybrids are formed when the discrete inorganic building blocks, e.g. clusters, are covalently bonded to the organic polymers or inorganic and organic polymers are covalently connected with each other.

Distinction between Nanocomposites and Hybrid Materials

The term nanocomposite is used if one of the structural units, either the organic or the inorganic, is in a defined size range of 1–100 nm. Large molecular building blocks for hybrid materials, such as large inorganic clusters, may be of the nanometer length scale. The term nanocomposites is generally used if discrete structural units in the respective size regime are used and the term hybrid materials is more often used if the inorganic units are formed in situ by molecular precursors, for example applying sol–gel reactions. Examples of discrete inorganic units for nanocomposites are nanoparticles, nanorods and carbon nanotubes. The biggest distinction between a nanocomposite and a hybrid is that a hybrid material possesses a property that does not exist in either of the parent components.

Advantages of Hybrid Materials over Traditional Composites

- Inorganic clusters or nanoparticles with specific optical, electronic or magnetic properties can be incorporated in organic polymer matrices.
- Contrary to pure solid state inorganic materials that often require a high temperature treatment for their processing, hybrid materials show a more polymer-like handling, either because of their large organic content or because of the formation of crosslinked inorganic networks from small molecular precursors just like in polymerization reactions.
- Light scattering in homogeneous hybrid material can be avoided and therefore optical transparency of the resulting hybrid materials and nanocomposites can be achieved.

Synthesis

Two different approaches can be used for the formation of hybrid materials: Either well-defined preformed building blocks are applied that react with each other to form the final hybrid material in which the precursors still at least partially keep their original integrity or one or both structural units are formed from the precursors that are transformed into a new (network) structure. It is important that the interface between the inorganic and organic materials which has to be tailored to overcome serious problems in the preparation of hybrid materials. Different building blocks and approaches can be used for their preparation and these have to be adapted to bridge the differences of inorganic and organic materials.

Building Block Approach

Building blocks at least partially keep their molecular integrity throughout the material formation, which means that structural units that are present in these sources for materials formation can also be found in the final material. At the same time typical properties of these building blocks usually survive the matrix formation, which is not the case if material precursors are transferred into novel materials. Representative examples of such well-defined building blocks are modified inorganic clusters or nanoparticles with attached reactive organic groups.

Cluster compounds often consist of at least one functional group that allows an interaction with an organic matrix, for example by copolymerization. Depending on the number of groups that can interact, these building blocks are able to modify an organic matrix (one functional group) or form partially or fully crosslinked materials (more than one group). For instance, two reactive groups can lead to the formation of chain structures. If the building blocks contain at least three reactive groups they can be used without additional molecules for the formation of a crosslinked material.

Beside the molecular building blocks mentioned, nanosized building blocks, such as particles or nanorods, can also be used to form nanocomposites. The building block approach has one large advantage compared with the in situ formation of the inorganic or organic entities: because at least one structural unit (the building block) is well-defined and usually does not undergo significant structural changes during the matrix formation, better structure–property predictions are possible. Furthermore, the building blocks can be designed in such a way to give the best performance in the materials' formation, for example good solubility of inorganic compounds in organic monomers by surface groups showing a similar polarity as the monomers.

In recent years many building blocks have been synthesized and used for the preparation of hybrid materials. Chemists can design these compounds on a molecular scale with highly sophisticated methods and the resulting systems are used for the formation of functional hybrid materials. Many future applications, in particular in nanotechnology, focus on a bottom-up approach in which complex structures are hierarchically formed by these small building blocks. This idea is also one of the driving forces of the building block approach in hybrid materials.

In Situ Formation of The Components

The in situ formation of the hybrid materials is based on the chemical transformation of the precursors used throughout materials' preparation. Typically this is the case if organic polymers are formed but also if the sol–gel process is applied to produce the inorganic component. In these cases well-defined discrete molecules are transformed to multidimensional structures, which often show totally different properties from the original precursors. Generally simple, commercially available molecules are applied and the internal structure of the final material is determined by the composition of these precursors but also by the reaction conditions. Therefore control over the latter is a crucial step in this process. Changing one parameter can often lead to two very different materials. If, for example, the inorganic species is a silica derivative formed by the sol–gel process, the change from base to acid catalysis makes a large difference because base catalysis leads to a more particle-like microstructure while acid catalysis leads to a polymer-like microstructure. Hence, the final performance of the derived materials is strongly dependent on their processing and its optimization.

In Situ Formation of Inorganic Materials

Many of the classical inorganic solid state materials are formed using solid precursors and high temperature processes, which are often not compatible with the presence of organic groups because they are decomposed at elevated temperatures. Hence, these high temperature processes are not suitable for the in situ formation of hybrid materials. Reactions that are employed should have more the character of classical covalent bond formation in solutions. One of the most prominent processes which fulfill these demands is the sol–gel process. However, such rather low temperature processes often do not lead to the thermodynamically most stable structure but to kinetic products, which has some implications for the structures obtained. For example low temperature derived inorganic materials are often amorphous or crystallinity is only observed on a very small length scale, i.e. the nanometer range. An example of the latter is the formation of metal nanoparticles in organic or inorganic matrices by reduction of metal salts or organometallic precursors.

Some methods of in situ formation of inorganic materials are:

- Sol-gel process.
- Nonhydrolytic sol–gel process.
- Sol–gel reactions of non-silicates.

Formation of Organic Polymers in Presence of Preformed Inorganic Materials

If the organic polymerization occurs in the presence of an inorganic material to form the hybrid material one has to distinguish between several possibilities to overcome the incompatibility of the two species. The inorganic material can either have no surface functionalization but the bare material surface; it can be modified with nonreactive organic groups (e.g. alkyl chains); or it can contain reactive surface groups such as polymerizable functionalities. Depending on these prerequisites the material can be pretreated, for example a pure inorganic surface can be treated with surfactants or silane coupling agents to make it compatible with the organic monomers, or functional monomers can be added that react with the surface of the inorganic material. If the inorganic component has nonreactive organic groups attached to its surface and it can be dissolved in a monomer which is subsequently polymerized, the resulting material after the organic polymerization, is a blend. In this case the inorganic component interact only weakly or not at all with the organic polymer; hence, a class I material is formed. Homogeneous materials are only obtained in this case if agglomeration of the inorganic components in the organic environment is prevented. This can be achieved if the interactions between the inorganic components and the monomers are better or at least the same as between the inorganic components. However, if no strong chemical interactions are formed, the long-term stability of a once homogeneous material is questionable because of diffusion effects in the resulting hybrid material. The stronger the respective interaction between the components, the more stable is the final material. The strongest interaction is achieved if class II materials are formed, for example with covalent interactions.

Hybrid Materials by Simultaneous Formation of both Components

Simultaneous formation of the inorganic and organic polymers can result in the most homogeneous type of interpenetrating networks. Usually the precursors for the sol–gel process are mixed with monomers for the organic polymerization and both processes are carried out at the same time with or without solvent. Applying this method, three processes are competing with each other:

- the kinetics of the hydrolysis and condensation forming the inorganic phase,
- the kinetics of the polymerization of the organic phase,
- the thermodynamics of the phase separation between the two phases.

By tailoring the kinetics of the two polymerizations in such a way that they occur simultaneously and rapidly enough, phase separation is avoided or minimized. Additional parameters such as attractive interactions between the two moieties, as described above can also be used to avoid phase separation.

One problem that also arises from the simultaneous formation of both networks is the sensitivity of many organic polymerization processes for sol–gel conditions or the composition of the materials formed. Ionic polymerizations, for example, often interact with the precursors or intermediates formed in the sol–gel process. Therefore, they are not usually applied in these reactions.

Applications

- Decorative coatings obtained by the embedding of organic dyes in hybrid coatings.
- Scratch-resistant coatings with hydrophobic or anti-fogging properties.
- Nanocomposite based devices for electronic and optoelectronic applications including light-emitting diodes, photodiodes, solar cells, gas sensors and field effect transistors.
- Fire retardant materials for construction industry.
- Nanocomposite based dental filling materials.
- Composite electrolyte materials for applications such as solid-state lithium batteries or supercapacitors.
- Proton conducting membranes used in fuel cells.
- Antistatic / anti-reflection coatings.
- Corrosion protection.
- Porous hybrid materials.

Nanowire

A nanowire is a nanostructure, with the diameter of the order of a nanometer (10^{-9} meters). It can also be defined as the ratio of the length to width being greater than 1000. Alternatively, nanowires can be defined as structures that have a thickness or diameter constrained to tens of nanometers or less and an unconstrained length. At these scales, quantum mechanical effects are important – which coined the term "quantum wires". Many different types of nanowires exist, including superconducting (e.g., YBCO), metallic (e.g., Ni, Pt, Au), semiconducting (e.g., Si, InP, GaN, etc.), and insulating (e.g., SiO2, TiO2). Molecular nanowires are composed of repeating molecular units either organic (e.g. DNA) or inorganic (e.g. $Mo_6S_{9-x}I_x$).

Overview

Typical nanowires exhibit aspect ratios (length-to-width ratio) of 1000 or more. As such they are often referred to as one-dimensional (1-D) materials. Nanowires have many interesting properties that are not seen in bulk or 3-D (three-dimensional) materials. This is because electrons in nanowires are quantum confined laterally and thus occupy energy levels that are different from the traditional continuum of energy levels or bands found in bulk materials.

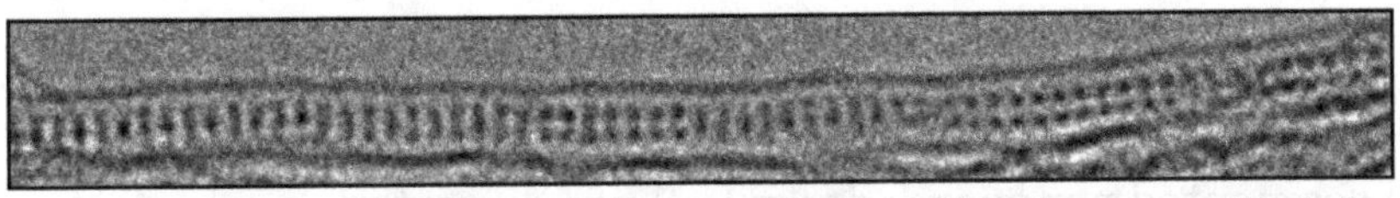

Crystalline 2×2-atom tin selenide nanowire grown inside a single-wall carbon nanotube (tube diameter ~1 nm).

Peculiar features of this quantum confinement exhibited by certain nanowires manifest themselves in discrete values of the electrical conductance. Such discrete values arise from a quantum mechanical restraint on the number of electrons that can travel through the wire at the nanometer scale. These discrete values are often referred to as the quantum of conductance and are integer multiples of

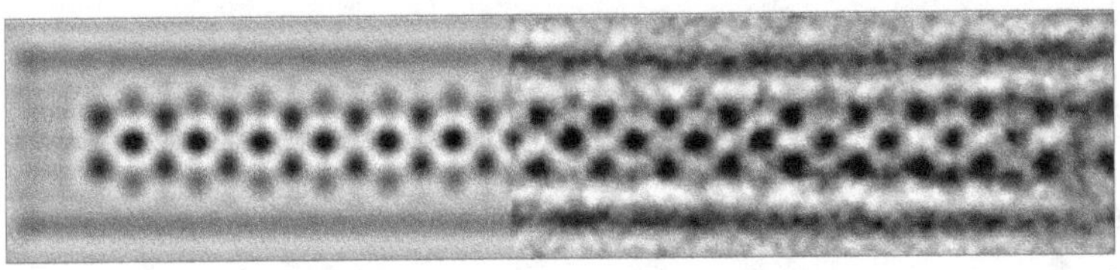

A noise-filtered HRTEM image of a HgTe extreme nanowire embedded down the central pore of a SWCNT. The image is also accompanied by a simulation of the crystal structure.

They are inverse of the well-known resistance unit h/e^2, which is roughly equal to 25812.8 ohms, and referred to as the von Klitzing constant R_K (after Klaus von Klitzing, the discoverer of exact quantization). Since 1990, a fixed conventional value $R_{K\text{-}90}$ is accepted.

Examples of nanowires include inorganic molecular nanowires ($Mo_6S_{9-x}I_x$, $Li_2Mo_6Se_6$), which can have a diameter of 0.9 nm and be hundreds of micrometers long. Other important examples are based on semiconductors such as InP, Si, GaN, etc., dielectrics (e.g. SiO_2,TiO_2), or metals (e.g. Ni, Pt).

There are many applications where nanowires may become important in electronic, opto-electronic and nanoelectromechanical devices, as additives in advanced composites, for metallic interconnects in nanoscale quantum devices, as field-emitters and as leads for biomolecular nanosensors.

Synthesis of Nanowires

There are two basic approaches to synthesizing nanowires: top-down and bottom-up. A top-down approach reduces a large piece of material to small pieces, by various means such as lithography or electrophoresis. A bottom-up approach synthesizes the nanowire by combining constituent adatoms. Most synthesis techniques use a bottom-up approach.

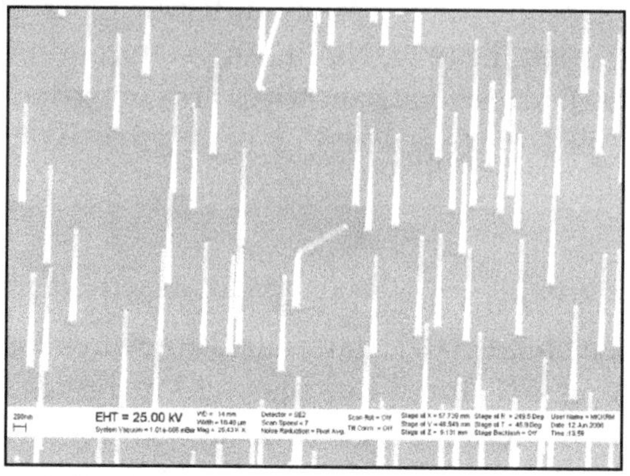

An SEM image of epitaxial nanowire heterostructures grown from catalytic gold nanoparticles.

Nanowire production uses several common laboratory techniques, including suspension, electrochemical deposition, vapor deposition, and VLS growth. Ion track technology enables growing homogeneous and segmented nanowires down to 8 nm diameter.

Suspension

A suspended nanowire is a wire produced in a high-vacuum chamber held at the longitudinal extremities. Suspended nanowires can be produced by:

- The chemical etching of a larger wire.
- The bombardment of a larger wire, typically with highly energetic ions.
- Indenting the tip of a STM in the surface of a metal near its melting point, and then retracting it.

VLS Growth

A common technique for creating a nanowire is vapor-liquid-solid method (VLS). This process can produce crystalline nanowires of some semiconductor materials. It uses a source material from either laser ablated particles or a feed gas such as silane.

VLS synthesis requires a catalyst. For nanowires, the best catalysts are liquid metal (such as gold) nanoclusters, which can either be self-assembled from a thin film by dewetting, or purchased in colloidal form and deposited on a substrate.

The source enters these nanoclusters and begins to saturate them. On reaching supersaturation, the source solidifies and grows outward from the nanocluster. Simply turning off the source can adjust the final length of the nanowire. Switching sources while still in the growth phase can create compound nanowires with super-lattices of alternating materials.

A single-step vapour phase reaction at elevated temperature synthesises inorganic nanowires such as $Mo_6S_{9-x}I_x$. From another point of view, such nanowires are cluster polymers.

Solution-phase Synthesis

Solution-phase synthesis refers to techniques that grow nanowires in solution. They can produce nanowires of many types of materials. Solution-phase synthesis has the advantage that it can produce very large quantities, compared to other methods. In one technique, the polyol synthesis, ethylene glycol is both solvent and reducing agent. This technique is particularly versatile at producing nanowires of lead, platinum, and silver.

The supercritical fluid-liquid-solid growth method can be used to synthesize semiconductor nanowires, e.g., Si and Ge. By using metal nanocrystals as seeds, Si and Ge organometallic precursors are fed into a reactor filled with a supercritical organic solvent, such as toluene. Thermolysis results in degradation of the precursor, allowing release of Si or Ge, and dissolution into the metal nanocrystals. As more of the semiconductor solute is added from the supercritical phase (due to a concentration gradient), a solid crystallite precipitates, and a nanowire grows uniaxially from the nanocrystal seed.

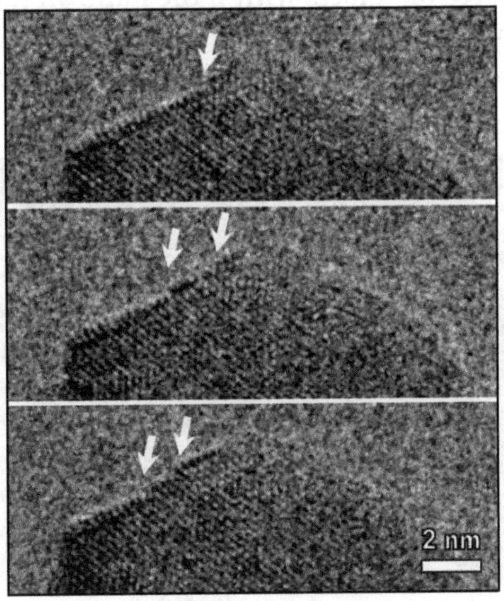

In situ observation of CuO nanowire growth.

Non-catalytic Growth

Nanowires can be also grown without the help of catalysts, which gives an advantage of pure nanowires and minimizes the number of technological steps. The simplest methods to obtain metal oxide nanowires use ordinary heating of the metals, e.g. metal wire heated with battery, by Joule heating in air can be easily done at home. The vast majority of nanowire-formation mechanisms are explained through the use of catalytic nanoparticles, which drive the nanowire growth and are either added intentionally or generated during the growth. However the mechanisms for catalyst-free growth of nanowires (or whiskers) were known from 1950s. Spontaneous nanowire formation by non-catalytic methods were explained by the dislocation present in specific directions or the growth anisotropy of various crystal faces. More recently, after microscopy advancement, the nanowire growth driven by screw dislocations or twin boundaries were demonstrated. The picture on the right shows a single atomic layer growth on the tip of CuO nanowire, observed by in situ TEM microscopy during the non-catalytic synthesis of nanowire.

Physics of Nanowires

Conductivity of Nanowires

Several physical reasons predict that the conductivity of a nanowire will be much less than that of the corresponding bulk material. First, there is scattering from the wire boundaries, whose effect will be very significant whenever the wire width is below the free electron mean free path of the bulk material. In copper, for example, the mean free path is 40 nm. Copper nanowires less than 40 nm wide will shorten the mean free path to the wire width.

Nanowires also show other peculiar electrical properties due to their size. Unlike single wall carbon nanotubes, whose motion of electrons can fall under the regime of ballistic transport (meaning the electrons can travel freely from one electrode to the other), nanowire conductivity is strongly influenced by edge effects. The edge effects come from atoms that lay at the nanowire surface and are not fully bonded to neighboring atoms like the atoms within the bulk of the nanowire. The unbonded atoms are often a source of defects within the nanowire, and may cause the nanowire to conduct electricity more poorly than the bulk material. As a nanowire shrinks in size, the surface atoms become more numerous compared to the atoms within the nanowire, and edge effects become more important.

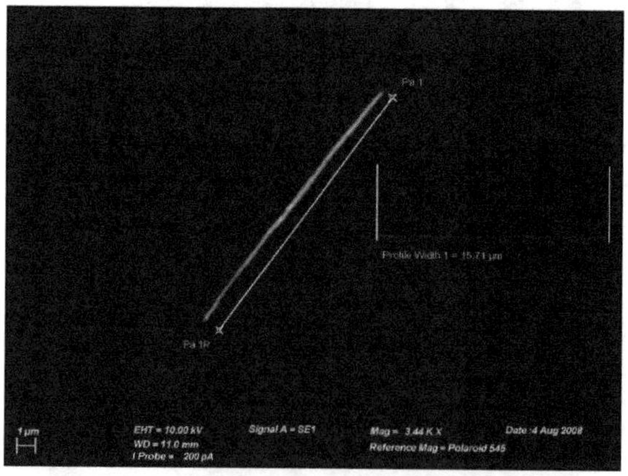

An SEM image of a 15 micrometer nickel wire.

Furthermore, the conductivity can undergo a quantization in energy: i.e. the energy of the electrons going through a nanowire can assume only discrete values, which are multiples of the conductance quantum $G = 2e^2/h$ (where e is the charge of the electron and h is the Planck constant.

The conductivity is hence described as the sum of the transport by separate *channels* of different quantized energy levels. The thinner the wire is, the smaller the number of channels available to the transport of electrons.

This quantization has been demonstrated by measuring the conductivity of a nanowire suspended between two electrodes while pulling it: as its diameter reduces, its conductivity decreases in a stepwise fashion and the plateaus correspond to multiples of G.

The quantization of conductivity is more pronounced in semiconductors like Si or GaAs than in metals, due to their lower electron density and lower effective mass. It can be observed in 25 nm wide silicon fins, and results in increased threshold voltage. In practical terms, this means that a MOSFET with such nanoscale silicon fins, when used in digital applications, will need a higher gate (control) voltage to switch the transistor on.

Welding Nanowires

To incorporate nanowire technology into industrial applications, researchers in 2008 developed a method of welding nanowires together: a sacrificial metal nanowire is placed adjacent to the ends of the pieces to be joined (using the manipulators of a scanning electron microscope); then an electric current is applied, which fuses the wire ends. The technique fuses wires as small as 10 nm.

For nanowires with diameters less than 10 nm, existing welding techniques, which require precise control of the heating mechanism and which may introduce the possibility of damage, will not be practical. Recently scientists discovered that single-crystalline ultrathin gold nanowires with diameters ~3–10 nm can be "cold-welded" together within seconds by mechanical contact alone, and under remarkably low applied pressures (unlike macro- and micro-scale cold welding process). High-resolution transmission electron microscopy and in situ measurements reveal that the welds are nearly perfect, with the same crystal orientation, strength and electrical conductivity as the rest of the nanowire. The high quality of the welds is attributed to the nanoscale sample dimensions, oriented-attachment mechanisms and mechanically assisted fast surface diffusion. Nanowire welds were also demonstrated between gold and silver, and silver nanowires (with diameters ~5–15 nm) at near room temperature, indicating that this technique may be generally applicable for ultrathin metallic nanowires. Combined with other nano- and microfabrication technologies, cold welding is anticipated to have potential applications in the future bottom-up assembly of metallic one-dimensional nanostructures.

Applications of Nanowires

Electronic Devices

Nanowires can be used for transistors. Transistors are used widely as fundamental building element in today's electronic circuits. As predicted by Moore's law, the dimension of transistors is shrinking smaller and smaller into nanoscale. One of the key challenges of building future nanoscale transistors is ensuring good gate control over the channel. Due to the high aspect ratio, if the gate dielectric is wrapped around the nanowire channel, we can get good control of channel electrostatic potential, thereby turning the transistor on and off efficiently.

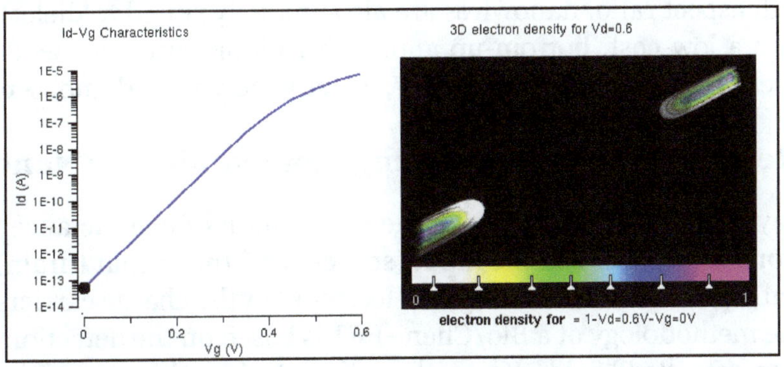

Atomistic simulation result for formation of inversion channel (electron density) and attainment of threshold voltage (IV) in a nanowire MOSFET. Note that the threshold voltage for this device lies around 0.45V.

To create active electronic elements, the first key step was to chemically dope a semiconductor nanowire. This has already been done to individual nanowires to create p-type and n-type semiconductors.

The next step was to find a way to create a p–n junction, one of the simplest electronic devices. This was achieved in two ways. The first way was to physically cross a p-type wire over an n-type wire. The second method involved chemically doping a single wire with different dopants along the length. This method created a p-n junction with only one wire.

After p-n junctions were built with nanowires, the next logical step was to build logic gates. By connecting several p-n junctions together, researchers have been able to create the basis of all logic circuits: the AND, OR, and NOT gates have all been built from semiconductor nanowire crossings.

In August 2012, researchers reported constructing the first NAND gate from undoped silicon nanowires. This avoids the problem of how to achieve precision doping of complementary nanocircuits, which is unsolved. They were able to control the Schottky barrier to achieve low-resistance contacts by placing a silicide layer in the metal-silicon interface.

It is possible that semiconductor nanowire crossings will be important to the future of digital computing. Though there are other uses for nanowires beyond these, the only ones that actually take advantage of physics in the nanometer regime are electronic.

In addition, nanowires are also being studied for use as photon ballistic waveguides as interconnects in quantum dot/quantum effect well photon logic arrays. Photons travel inside the tube, electrons travel on the outside shell.

When two nanowires acting as photon waveguides cross each other the juncture acts as a quantum dot.

Conducting nanowires offer the possibility of connecting molecular-scale entities in a molecular computer. Dispersions of conducting nanowires in different polymers are being investigated for use as transparent electrodes for flexible flat-screen displays.

Because of their high Young's moduli, their use in mechanically enhancing composites is being investigated. Because nanowires appear in bundles, they may be used as tribological additives to improve friction characteristics and reliability of electronic transducers and actuators.

Because of their high aspect ratio, nanowires are also uniquely suited to dielectrophoretic manipulation, which offers a low-cost, bottom-up approach to integrating suspended dielectric metal oxide nanowires in electronic devices such as UV, water vapor, and ethanol sensors.

Sensing of Proteins and Chemicals Using Semiconductor Nanowires

In an analogous way to FET devices in which the modulation of conductance (flow of electrons/holes) in the semiconductor, between the input (source) and the output (drain) terminals, is controlled by electrostatic potential variation (gate-electrode) of the charge carriers in the device conduction channel, the methodology of a Bio/Chem-FET is based on the detection of the local change in charge density, or so-called "field effect", that characterizes the recognition event between a target molecule and the surface receptor.

This change in the surface potential influences the Chem-FET device exactly as a 'gate' voltage does, leading to a detectable and measurable change in the device conduction. When these devices are fabricated using semiconductor nanowires as the transistor element the binding of a chemical or biological species to the surface of the sensor can lead to the depletion or accumulation of charge carriers in the "bulk" of the nanometer diameter nanowire i.e. (small cross section available for conduction channels). Moreover, the wire, which serves as a tunable conducting channel, is in close contact with the sensing environment of the target, leading to a short response time, along with orders of magnitude increase in the sensitivity of the device as a result of the huge S/V ratio of the nanowires.

While several inorganic semiconducting materials such as Si, Ge, or metal oxides (e.g. $In2O3$, $SnO2$, ZnO, etc.) have been used for the preparation of nanowires. Silicon nanowires are usually the material of choice when fabricating nanowire FET-based chemo/biosensors.

Several examples of the use of silicon nanowire sensing devices include the ultra sensitive, real-time sensing of biomarker proteins for cancer, detection of single virus particles, and the detection of nitro-aromatic explosive materials such as 2,4,6 Tri-nitrotoluene (TNT) in sensitives superior to these of canines. Silicon nanowires could also be used in their twisted form, as electromechanical devices, to measure intermolecular forces with great precision.

Limitations of Sensing With Silicon Nanowire FET Devices

Generally, the charges on dissolved molecules and macromolecules are screened by dissolved counterions, since in most cases molecules bound to the devices are separated from the sensor surface by approximately 2–12 nm (the size of the receptor proteins or DNA linkers bound to the sensor surface). As a result of the screening, the electrostatic potential that arises from charges on the analyte molecule decays exponentially toward zero with distance. Thus, for optimal sensing, the Debye length must be carefully selected for nanowire FET measurements. One approach of overcoming this limitation employs fragmentation of the antibody-capturing units and control over surface receptor density, allowing more intimate binding to the nanowire of the target protein. This approach proved useful for dramatically enhancing the sensitivity of cardiac biomarkers (e.g. Troponin) detection directly from serum for the diagnosis of acute myocardial infarction.

Nanocellulose

Nanocellulose is a term referring to nano-structured cellulose. This may be either cellulose nanofibers (CNF) also called microfibrillated cellulose (MFC), nanocrystalline cellulose (NCC), or bacterial nanocellulose, which refers to nano-structured cellulose produced by bacteria.

CNF is a material composed of nanosized cellulose fibrils with a high aspect ratio (length to width ratio). Typical lateral dimensions are 5–20 nanometers and longitudinal dimension is in a wide range, typically several micrometers. It is pseudo-plastic and exhibits the property of certain gels or fluids that are thick (viscous) under normal conditions, but flow (become thin, less viscous) over time when shaken, agitated, or otherwise stressed. This property is known as thixotropy. When the shearing forces are removed the gel regains much of its original state. The fibrils are isolated from any cellulose containing source including wood-based fibers (pulp fibers) through high-pressure, high temperature and high velocity impact homogenization, grinding or microfluidization.

Nanocellulose

Nanocellulose can also be obtained from native fibers by an acid hydrolysis, giving rise to highly crystalline and rigid nanoparticles (often referred to as CNC or nanowhiskers) which are shorter (100s to 1000 nanometers) than the nanofibrils obtained through the homogenization, microfluiodization or grinding routes. The resulting material is known as nanocrystalline cellulose (NCC).

History and Terminology

The terminology microfibrillated/nanocellulose or (MFC) was first used by Turbak, Snyder and Sandberg in the late 1970s at the ITT Rayonier labs in Whippany, New Jersey, USA to describe a product prepared as a gel type material by passing wood pulp through a Gaulin type milk homogenizer at high temperatures and high pressures followed by ejection impact against a hard surface. The terminology (MFC) first appeared publicly in the early 1980s when a number of patents and publications were issued to ITT Rayonier on this totally new nanocellulose composition of matter. In later work Herrick at Rayonier also published work on making a dry powder form of the gel. Since Rayonier is one of the world's premier producers of purified pulps their business interests have always been 1) to create new uses and new markets for pulps and 2) never to compete with new or potentially new customers. Thus, as the patents issued, Rayonier gave free license to whoever wanted to pursue this new use for cellulose. Rayonier,as a company, never pursued scale-up. Rather, Turbak et al. pursued 1) finding new uses for the MFC/nanocellulose. These included using MFC as a thickener and binder in foods, cosmetics, paper formation, textiles, nonwovens, etc. and 2) evaluate swelling and other techniques for lowering the energy requirements for MFC/ Nanocellulose production. ITT closed the Rayonier Whippany Labs in 1983–84 and further work on making a dry powder form of MFC was done by Herric at the Rayonier labs in Shelton, Washington, USA.

The field was later taken up in Japan in the mid 1990s by the group of Taniguchi and co-workers and later by Yano and co-workers. and a host of major companies (see numerous U.S. patents issued to P&G, J&J, 3M, McNeil, etc. using U.S. patent search under inventor name Turbak search base). Today, there are still extensive research and development efforts around the world in this field.

As of August 2012, NCC was produced in a factory operated by CelluForce, producing one ton per day, and by a just-opened facility operated by the US Forest Service. Recent developments (as of January 2016) show a renewed effort to industrialise MFC and find new applications fields with newly founded companies such as Borregaard's Exilva, FiberLean (a joint-venture of Imerys and Omya), American Process and WEIDMANN Fiber Technology among others.

Manufacture

Nanocellulose/CNF or NCC can be prepared from any cellulose source material, but woodpulp is normally used.

The nanocellulose fibrils may be isolated from the wood-based fibers using mechanical methods which expose the pulp to high shear forces, ripping the larger wood-fibres apart into nanofibers. For this purpose high-pressure homogenizers, ultrasonic homogenizers, grinders or microfluidizers can be used. The homogenizers are used to delaminate the cell walls of the fibers and liberate the nanosized fibrils. This process is responsible for the high energy consumptions associated with the fiber delamination. Values over 30 MWh/tonne are not uncommon.

Pre-treatments are sometimes used to address this problem. Examples of such pre-treatments are enzymatic/mechanical pre-treatment and introduction of charged groups e.g. through carboxymethylation or TEMPO-mediated oxidation. It has been shown that energy consumption can be heavily decreased using these pre-treatments and Lindström and Ankerfors has reported values below 1 MWh/tonne.

Cellulose nanowhiskers, a more crystalline form of nanocellulose, are formed by the acid hydrolysis of native cellulose fibers commonly using sulfuric or hydrochloric acid. The amorphous sections of native cellulose are hydrolysed and after careful timing, the crystalline sections can be retrieved from the acid solution by centrifugation and washing. Cellulose nanowhiskers are rodlike highly crystalline particles (relative crystallinity index above 75%) with a rectangular cross section. Their dimensions depend on the native cellulose source material, and hydrolysis time and temperature.

Recent breakthroughs in nanocellulose production were announced April 2013 by R. Malcolm Brown Jr. PHD, a biology professor at the University of Texas at Austin, who presented his team's findings at an American Chemical Society conference in New Orleans.

Borregaard AS, a Norwegian biorefinery company based in Sarpsborg, Norway, announced in October 2014 that they are investing in a facility to produce microfibrillated cellulose on a commercial scale. The production facility will have a capacity of 1000 tonnes per year, and is scheduled to be operational from Q3, 2016.

American Process Inc, a biorefinery technology development company headquartered in Atlanta, Georgia, has also recently announced a manufacturing process using sulfur dioxide and ethanol pretreatment of biomass to produce both cellulose nanofibrils and cellulose nanocrystals and lignin-coated, hydrophobic varieties of each.

ICAR-CIRCOT Nanocellulose Pilot Plant, a unique facility for production of nanocellulose from cotton and other agro-biomass, was inaugurated by Padma Visbhushan Dr. R. A. Mashelkar in Mumbai, India on August 21, 2015. This pilot plant was funded by World Bank supported National Agricultural Innovation Project of Indian Council of Agricultural Research, New Delhi. This pilot plant has the capacity of 10 kg nanocellulose production per shift of 8 hour operation and uses patented energy efficient and eco-friendly technologies. Both nanocrystalline cellulose and nanofibrillated cellulose can be produced in this plant (both in aqueous and dry form). This is established to act as an incubation center for new entrepreneurs and as technology demonstration unit.

Structure and Properties

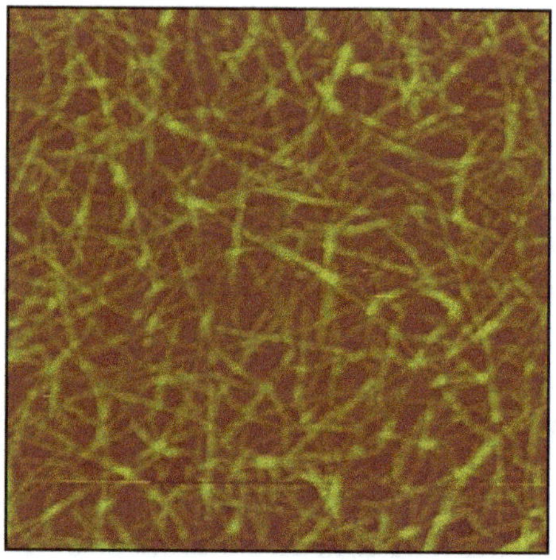

AFM height image of carboxymethylated nanocellulose adsorbed on a silica surface. The scanned surface area is 1μm2.

Nanocellulose Dimensions and Crystallinity

The ultrastructure of cellulose derived from various sources has been extensively studied. Techniques such as transmission electron microscopy (TEM), scanning electron microscopy (SEM), atomic force microscopy (AFM), wide angle X-ray scattering (WAXS), small incidence angle X-ray diffraction and solid state ^{13}C cross-polarization magic angle spinning (CP/MAS) nuclear magnetic resonance (NMR) spectroscopy have been used to characterize nanocellulose morphology. These methods have typically been applied for the investigation of dried nanocellulose morphology.

Although a combination of microscopic techniques with image analysis can provide information on nanocellulose fibril widths, it is more difficult to determine nanocellulose fibril lengths because of entanglements and difficulties in identifying both ends of individual nanofibrils. It has been reported that nanocellulose suspensions may not be homogeneous and that they consist of various structural components, including cellulose nanofibrils and nanofibril bundles.

Most methods have typically been applied to investigation of dried nanocellulose dimensions, although a study was conducted where the size and size-distribution of enzymatically pre-treated nanocellulose fibrils in a suspension was studied using cryo-TEM. The fibrils were found to be rather mono-dispersed mostly with a diameter of ca. 5 nm although occasionally thicker fibril bundles were present. It should be noted that, some newly published results indicated that by combining ultrasonication with an "oxidation pretreatment", cellulose microfibrils with a lateral dimension that belows 1 nm is observed by AFM. The lower end of the thickness dimension is around 0.4 nm, which is related to the thickness of a cellulose monolayer sheet.

The aggregate widths can be determined by CP/MAS NMR developed by Innventia AB, Sweden, which also has been demonstrated to work for nanocellulose (enzymatic pre-treatment). An average width of 17 nm has been measured with the NMR-method, which corresponds well with SEM and TEM. Using TEM, values of 15 nm have been reported for nanocellulose from carboxymethylated pulp. However, also thinner fibrils can also be detected. Wågberg et al. reported fibril widths of 5–15 nm for a nanocellulose with a charge density of about 0.5 meq./g. The group of Isogai reported fibril widths of 3–5 nm for TEMPO-oxidized cellulose having a charge density of 1.5 meq./g.

The influence of cellulose pulp chemistry on the nanocellulose microstructure has been investigated using AFM to compare the microstructure of two types of nanocellulose prepared at Innventia AB (enzymatically pre-treated nanocellulose and carboxymethylated nanocellulose). Due to the chemistry involved in producing carboxymethylated nanocellulose, it differs significantly from the enzymatically pre-treated one. The number of charged groups on the fibril surfaces are very different. The carboxymethylation pre-treatment makes the fibrils highly charged and, hence, easier to liberate, which results in smaller and more uniform fibril widths (5–15 nm) compared to the enzymatically pre-treated nanocellulose, where the fibril widths were 10–30 nm. The degree of crystallinity and the cellulose crystal structure of nanocellulose were also studied at the same time. The results clearly showed the nanocellulose exhibited cellulose crystal I organization and that the degree of crystallinity was unchanged by the preparation of the nanocellulose. Typical values for the degree of crystallinity were around 63%.

Viscosity

The unique rheology of nanocellulose dispersions was recognized by the early investigators. The high viscosity at low nanocellulose concentrations makes nanocellulose very interesting as a non-caloric stabilizer and gellant in food applications, the major field explored by the early investigators.

The dynamic rheological properties were investigated in great detail and revealed that the storage and loss modulus were independent of the angular frequency at all nanocellulose concentrations between 0.125% to 5.9%. The storage modulus values are particularly high (104 Pa at 3% concentration) compared to results for cellulose nanowhiskers (102 Pa at 3% concentration). There is also a particular strong concentration dependence as the storage modulus increases 5 orders of magnitude if the concentration is increased from 0.125% to 5.9%.

Nanocellulose gels are also highly shear thinning (the viscosity is lost upon introduction of the shear forces). The shear-thinning behaviour is particularly useful in a range of different coating applications.

Mechanical Properties

Crystalline cellulose has interesting mechanical properties for use in material applications. Its tensile strength is about 500MPa, similar to that of aluminium. Its stiffness is about 140–220 GPa, comparable with that of Kevlar and better than that of glass fiber, both of which are used commercially to reinforce plastics. Films made from nanocellulose have high strength (over 200 MPa), high stiffness (around 20 GPa) and high strain (12%). Its strength/weight ratio is 8 times that of stainless steel.

Barrier Properties

In semi-crystalline polymers, the crystalline regions are considered to be gas impermeable. Due to relatively high crystallinity, in combination with the ability of the nanofibers to form a dense network held together by strong inter-fibrillar bonds (high cohesive energy density), it has been suggested that nanocellulose might act as a barrier material. Although the number of reported oxygen permeability values are limited, reports attribute high oxygen barrier properties to nanocellulose films. One study reported an oxygen permeability of 0.0006 (cm^3 µm)/(m^2 day kPa) for a ca. 5 µm thin nanocellulose film at 23 °C and 0% RH. In a related study, a more than 700-fold decrease in oxygen permeability of a polylactide (PLA) film when a nanocellulose layer was added to the PLA surface was reported.

The influence of nanocellulose film density and porosity on film oxygen permeability has recently been explored. Some authors have reported significant porosity in nanocellulose films, which seems to be in contradiction with high oxygen barrier properties, whereas Aulin et al. measured a nanocellulose film density close to density of crystalline cellulose (cellulose Iß crystal structure, 1.63 g/cm^3) indicating a very dense film with a porosity close to zero.

Changing the surface functionality of the cellulose nanoparticle can also affect the permeability of nanocellulose films. Films constituted of negatively charged cellulose nanowhiskers could effectively reduce permeation of negatively charged ions, while leaving neutral ions virtually unaffected. Positively charged ions were found to accumulate in the membrane.

Multi-Parametric Surface Plasmon Resonance is one of the methods to study barrier properties of natural, modified or coated nanocellulose. The different antifouling, moisture, solvent, antimicrobial barrier formulation quality can be measured on the nanoscale. The adsorption kinetics as well as the degree of swelling can be measured in real-time and label-free.

Foams

Nanocellulose can also be used to make aerogels/foams, either homogeneously or in composite formulations. Nanocellulose-based foams are being studied for packaging applications in order to replace polystyrene-based foams. Svagan et al. showed that nanocellulose has the ability to reinforce starch foams by using a freeze-drying technique. The advantage of using nanocellulose instead of wood-based pulp fibers is that the nanofibrills can reinforce the thin cells in the starch foam. Moreover, it is possible to prepare pure nanocellulose aerogels applying various freeze-drying and super critical CO2 drying techniques. Aerogels and foams can be used as porous templates. Tough ultra-high porosity foams prepared from cellulose I nanofibrill suspensions were studied by Sehaquiet al. A wide range of mechanical properties including compression was obtained by controlling density and nanofibrill interaction in the foams. Cellulose nanowhiskers could also be made to gel in water under low power sonication giving rise to aerogels with the highest reported surface area (>600m2/g) and lowest shrinkage during drying (6.5%) of cellulose aerogels. In another study by Aulin et al., the formation of structured porous aerogels of nanocellulose by freeze-drying was demonstrated. The density and surface texture of the aerogels was tuned by selecting the concentration of the nanocellulose dispersions before freeze-drying. Chemical vapour deposition of a fluorinated silane was used to uniformly coat the aerogel to tune their wetting properties towards non-polar liquids/oils. The authors demonstrated that it is possible to switch the wettability behaviour of the cellulose surfaces between super-wetting and super-repellent, using different scales of roughness and porosity created by the freeze-drying technique and change of concentration of the nanocellulose dispersion. Structured porous cellulose foams can however also be obtained by utilizing the freeze-drying technique on cellulose generated by Gluconobacter strains of bacteria, which bio-synthesize open porous networks of cellulose fibers with relatively large amounts of nanofibrills dispersed inside. Olsson et al. demonstrated that these networks can be further impregnated with metalhydroxide/oxide precursors, which can readily be transformed into grafted magnetic nanoparticles along the cellulose nanofibers. The magnetic cellulose foam may allow for a number of novel applications of nanocellulose and the first remotely actuated magnetic super sponges absorbing 1 gram of water within a 60 mg cellulose aerogel foam were reported. Notably, these highly porous foams (>98% air) can be compressed into strong magnetic nanopapers, which may find use as functional membranes in various applications.

Surface Modification

The surface modification of nanocellulose is currently receiving a large amount of attention. Nanocellulose displays a high concentration of hydroxyl groups at the surface which can be reacted. However, hydrogen bonding strongly affects the reactivity of the surface hydroxyl groups. In addition, impurities at the surface of nanocellulose such as glucosidic and lignin fragments need to be removed before surface modification to obtain acceptable reproducibility between different batches.

Cellulose nanofibers can also be modified as cationic. This cationic cellulose increase the affinity for anions.

Safety Aspects

Health, safety and environmental aspects of nanocellulose have been recently evaluated. Processing of nanocellulose does not cause significant exposure to fine particles during friction grinding or spray drying. No evidence of inflammatory effects or cytotoxicity on mouse or human macrophages can be observed after exposure to nanocellulose. The results of toxicity studies suggest that nanocellulose is not cytotoxic and does not cause any effects on inflammatory system in macrophages. In addition, nanocellulose is not acutely toxic to Vibrio fischeri in environmentally relevant concentrations.

Applications

The properties of nanocellulose (e.g. mechanical properties, film-forming properties, viscosity etc.) makes it an interesting material for many applications and the potential for a multibillion-dollar industry.

Paper and Paperboard

There is potential of nanocellulose applications in the area of paper and paperboard manufacture. Nanocelluloses are expected to enhance the fiber-fiber bond strength and, hence, have a strong reinforcement effect on paper materials. Nanocellulose may be useful as a barrier in grease-proof type of papers and as a wet-end additive to enhance retention, dry and wet strength in commodity type of paper and board products.

Composite

As described above the properties of the nanocellulose makes an interesting material for reinforcing plastics. Nanocellulose has been reported to improve the mechanical properties of, for example, thermosetting resins, starch-based matrixes, soy protein, rubber latex, poly(lactide). The composite applications may be for use as coatings and films, paints, foams, packaging.

Food

Nanocellulose can be used as a low calorie replacement for today's carbohydrate additives used as thickeners, flavour carriers and suspension stabilizers in a wide variety of food products and is useful for producing fillings, crushes, chips, wafers, soups, gravies, puddings etc. The food applications were early recognised as a highly interesting application field for nanocellulose due to the rheological behaviour of the nanocellulose gel.

Hygiene and Absorbent Products

Different applications in this field include:

- Super water absorbent (e.g. for incontinence pads material).
- Nanocellulose used together with super absorbent polymers.
- Use of nanocellulose in tissue, non-woven products or absorbent structures.
- Use as antimicrobial films.

Emulsion and Dispersion

Apart from the numerous applications in the area of food additives, the general area of emulsion and dispersion applications in other fields has also received some attention. Oil in water applications were early recognized. The area of non-settling suspensions for pumping sand, coal as well as paints and drilling muds was also explored by the early investigators.

Oil Recovery

Hydrocarbon fracturing of oil-bearing formations is a potentially interesting and large-scale application. Nanocellulose has been suggested for use in oil recovery applications as a fracturing fluid. Drilling muds based on nanocellulose have also been suggested.

Medical, Cosmetic and Pharmaceutical

The use of nanocellulose in cosmetics and pharmaceuticals was also early recognized. A wide range of high-end applications have been suggested:

- Freeze-dried nanocellulose aerogels used in sanitary napkins, tampons, diapers or as wound dressing.

- The use of nanocellulose as a composite coating agent in cosmetics e.g. for hair, eyelashes, eyebrows or nails.
- A dry solid nanocellulose composition in the form of tablets for treating intestinal orders.
- Nanocellulose films for screening of biological compounds and nucleic acids encoding a biological compound.
- Filter medium partly based on nanocellulose for leukocyte free blood transfusion.
- A buccodental formulation, comprising nanocellulose and a polyhydroxylated organic compound.
- Powdered nanocellulose has also been suggested as an excipient in pharmaceutical compositions.
- nanocellulose in compositions of a photoreactive noxious substance purging agent.
- Elastic cryo-structured gels for potential biomedical and biotechnological application.
- Matrix for 3D cell culture.

Other Applications

- Activate the dissolution of cellulose in different solvents.
- Regenerated cellulose products, such as fibers films, cellulose derivatives.
- Tobacco filter additive.
- Organometallic modified nanocellulose in battery separators.
- Reinforcement of conductive materials.
- Loud-speaker membranes.
- High-flux membranes.
- Flexible electronic displays.
- Computer components.
- Capacitors.
- Lightweight body armour and ballistic glass.
- Corrosion inhibitors.

References

- Pacios Pujadó, Mercè (2012). Carbon Nanotubes as Platforms for Biosensors with Electrochemical and Electronic Transduction. Springer Heidelberg. pp. XX,208. doi:10.1007/978-3-642-31421-6. ISBN 978-3-642-31421-6. ISSN 2190-5053

- Herrick, F.W.; R.L. Casebier; J.K. Hamilton; K.R. Sandberg (1983). "Microfibrillated cellulose: morphology and accessibility". In A. Sarko (ed.). Proceedings of the Ninth Cellulose Conference. Applied Polymer Symposia, 37. New York City: Wiley. pp. 797–813. ISBN 0-471-88132-5
- Chinga-Carrasco, G., Miettinen, A., Luengo Hendriks, C.-L., Gamstedt, E.K., Kataja, M. (2011). Structural Characterisation of Kraft Pulp Fibres and Their Nanofibrillated Materials for Biodegradable Composite Applications. InTech. ISBN 978-953-307-352-1
- Kolosnjaj J, Szwarc H, Moussa F (2007). "Toxicity studies of carbon nanotubes". Adv Exp Med Biol. Advances in Experimental Medicine and Biology. 620: 181–204. doi:10.1007/978-0-387-76713-0_14. ISBN 978-0-387-76712-3. PMID 18217344
- Lalwani, G; Henslee, AM; Farshid, B; Parmar, P; Lin, L; Qin, YX; Kasper, FK; Mikos, AG; Sitharaman, B (September 2013). "Tungsten disulfide nanotubes reinforced biodegradable polymers for bone tissue engineering". Acta biomaterialia. 9 (9): 8365–73. doi:10.1016/j.actbio.2013.05.018. PMC 3732565. PMID 23727293
- Godovsky, D. Y. (2000). "Device Applications of Polymer-Nanocomposites". In Chang, J. Y. Biopolymers · PVA Hydrogels, Anionic Polymerisation Nanocomposites. Advances in Polymer Science. 153. pp. 163–205. doi:10.1007/3-540-46414-X_4. ISBN 978-3-540-67313-2
- Berglund, Lars (2005). "Cellulose-based nanocomposites". In A.K. Mohanty; M. Misra; L. Drzal (Eds). Natural fibers, biopolymers and biocomposites. Boca Raton, Florida: CRC Press. pp. 807–832. ISBN 978-0-8493-1741-5
- Gatti, Teresa; Vicentini, Nicola; Mba, Miriam; Menna, Enzo (2016-02-01). "Organic Functionalized Carbon Nanostructures for Functional Polymer-Based Nanocomposites". European Journal of Organic Chemistry. 2016 (6): 1071–1090. doi:10.1002/ejoc.201501411. ISSN 1099-0690
- Janeta, Mateusz; John, Łukasz; Ejfler, Jolanta; Szafert, Sławomir (2014-11-24). "High-Yield Synthesis of Amido-Functionalized Polyoctahedral Oligomeric Silsesquioxanes by Using Acyl Chlorides". Chemistry: A European Journal. 20 (48): 15966–15974. doi:10.1002/chem.201404153. ISSN 1521-3765
- Lalwani, Gaurav; Henslee, A. M.; Farshid, B; Parmar, P; Lin, L; Qin, Y. X.; Kasper, F. K.; Mikos, A. G.; Sitharaman, B (September 2013). "Tungsten disulfide nanotubes reinforced biodegradable polymers for bone tissue engineering". Acta Biomaterialia. 9 (9): 8365–8373. doi:10.1016/j.actbio.2013.05.018. PMC 3732565. PMID 23727293
- Lalwani, Gaurav; Henslee, Allan M.; Farshid, Behzad; Lin, Liangjun; Kasper, F. Kurtis; Yi-, Yi-Xian; Qin, Xian; Mikos, Antonios G.; Sitharaman, Balaji (2013). "Two-dimensional nanostructure-reinforced biodegradable polymeric nanocomposites for bone tissue engineering". Biomacromolecules. 14 (3): 900–909. doi:10.1021/bm301995s. PMC 3601907. PMID 23405887
- Tian, Zhiting; Hu, Han; Sun, Ying (2013). "A molecular dynamics study of effective thermal conductivity in nanocomposites". Int. J. Heat Mass Transfer. 61: 577–582. doi:10.1016/j.ijheatmasstransfer.2013.02.023
- Lindström, Tom; Mikael Ankerfors (2009). "NanoCellulose Developments in Scandinavia". 7th International Paper and Coating Chemistry Symposium (Preprint CD ed.). Hamilton, Ontario: McMaster University Engineering. ISBN 978-0-9812879-0-4

5

Applications of Composite Materials

Composite materials find application in a variety of fields. They have automobile applications, aerospace applications, marine applications, biomedical applications, sports applications and civil applications. This chapter closely examines these key applications of composites to provide an extensive understanding of the subject.

Composite materials also called composition materials or shortened to composites. Composite materials are materials made from two or more than two materials with considerably differ in physical and chemical properties, that when combined, make a material with appearances different from the individual components. Composites comprise strong load carrying material is known as reinforcement and weaker materials is known as matrix. Reinforcement provides stiffness and strength which helps to support structural load. Composite materials do not lose their respective identities but still relate their properties to the product causing from their mixture. The benefits of composite materials have their great stiffness and strength. There are in many cases, the reinforcement is stronger, tougher, harder and stiffer than the matrix. It finds application in automotive, aerospace, electronic equipment, sport goods, furniture, medical equipment & packaging industry.

Composite materials used as an industrial material for their outstanding resistance to chemicals and most forms of corrosion. This property of composite material conventionally important is hardly the only useful property. There are many important and useful properties are, low mass, low weight unequalled manufacturing and processing possibilities, complex material body are easily produced, appropriate to very small products and very large product, tooling cost is very low, satisfactory surface finish can be an integral feature. Composites have four to six times tensile strength as compare to steel or aluminium (depending on the reinforcements). Composites have less noise and lower vibration transmission than metals at the time of operations. Composite materials have torsional stiffness and impact properties. Composites have high fatigue strength, impact, environmental resistance and reduce maintenance, higher fatigue endurance limit (up to 60% of ultimate tensile strength.

Composites exhibit fire retardancy and good corrosion resistance. Composites have improved surfaces properties and readily incorporable integral decorative melamine is other characteristics of composites, low electrical conductivity and thermal expansion. Composite parts can eliminate joints and providing simplification and assembly design compared to non-composite metallic parts. Composite materials have high cost of material, long development time, manufacturing difficulties, low ductility, temperature limits, solvent or moister attack, hidden damages and damage susceptibility. Matrix used in composite materials is subject to environmental degradation and analysis is difficult, hot curing is necessary in many cases requiring special tooling, hot or cold curing takes time. Materials require refrigerated transport and storage and have limited shelf life.

Automobile Applications

In Automotive, composites are being considered to make safer with low weight and more fuel-efficient vehicles. The fibre-reinforced composite is composed of a high strength fiber (i.e. carbon or glass) in a matrix material (polymer and metals i.e. Al, Mg etc.) and it can provide magnify properties compared with the individual materials by themselves. Many components like seat, roof, steering wheel, hatch, dashboard, mats, energy absorber, interior and exterior panel, wheels, leaf spring, engine cover etc. are fabricated by composite materials. For example, reducing a car mass by 100kg saves about 0.7 litre fuel each 100km (directly and indirectly). For the lightweight composite structures manufacturing process, the technologies which are based on fiber reinforced thermoplastic materials can be integrated. The main motivators for the lightweight materials applications are weight savings and possible cost savings. Significant weight reductions with improved performance will mean less fuel consumption and CO_2 emissions. The transport industries are customer sensitive and currently, the customers are pushing for the cost-effectiveness and the more environmentally friendly transport system. By using low cost, eco-friendly, and reliable materials the economic burden would be reduced for both the customer and the automotive industry. The demand for better fuel efficiency and reduced emissions developed the demand for weight reduction in order to comply with EU legislation (from $<130g\ CO_2/km$ in 2015 down to $<95g\ CO_2/km$ by 2021). Composites can offer lightweight benefits from 15-25% for glass-fiber reinforced composites (GFRP) to 25-40% for carbon-fiber reinforced composites (CFRP) in comparison to other structural metallic materials that are presently dominant, such as steel, iron, and aluminium.

Obstacles to use Composites in Automotive

Industry inexperience with polymer and Al-based composites materials, low production rate due to undeveloped processes, the new techniques need for joining, lack of knowledge about material responses to automotive environments, immature recycling technologies, a small supplier base and lack of crash models; these are the main obstacles for an automotive industry implementation of composites. Due to the high current cost of carbon fiber as compared to other structural materials for a vehicle is the additional factor for the carbon fiber based composites are restricted in industry use. By weight, about 8% of today's automobile parts are made of composites including bumpers, body panels, and doors. Several research and development efforts are needed for increasing the application of lightweight composites in the automotive sector. The relation between the cost reduction and weight reduction in automotive is shown in figure.

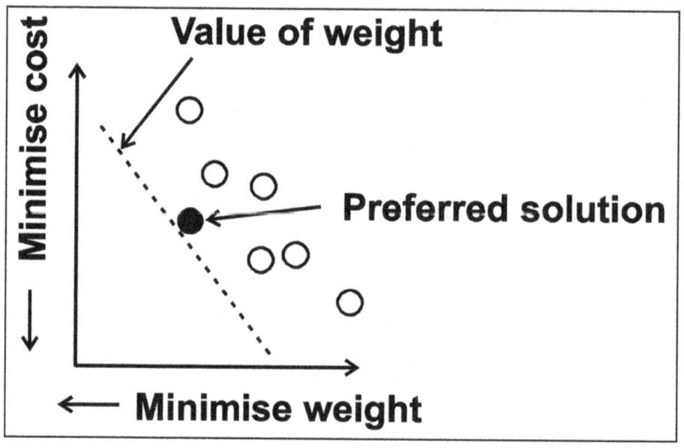

Weight reduction is inversely proportional to cost reduction.

Polymer Matrix Composites for Automotive

Polymer matrix reinforced with fiber or whiskers is the most common form of composite that has been used in the automotive industry. The first application of such materials was the fiber-glass body of the Chevrolet Corvette in 1953. The key advantages of polymer materials over the conventional metallic materials are their specific strength properties with weight saving of 20-40%, lower thermal expansion properties, potential for rapid process cycles, ability to meet stringent dimensional stability and excellent fatigue and fracture resistance. Automobile's segment of composites accounts for about 50% of the thermoplastic and 24% of the thermoset composite market in the world. For weight reduction, the glass fiber reinforced thermoplastic polymer is a promising material due to its relatively low cost, fast cycle time and ability to facilitate parts integration. Carbon fiber reinforced polymer is another composite but will require breakthroughs in the cost and manufacturing techniques to be cost-effective for high volume production. For the automotive sector, it is a favorable option for reductions in energy uses and lower emissions levels by making advanced composites. In automotive sector future business opportunities are as follows:

- RTM panel.
- Glass fiber/epoxy springs for heavy trucks or trailers.
- Valve guides.
- Rocker arm covers, wheels, and engine shrouds, suspension arms.
- Filament-wound fuel tanks.
- Electrical vehicle body components and assembly units.
- Automotive racing brakes and train brakes.
- Clutch plates.

A polymer composite body has significant implications for vehicle light-weighting and improved fuel efficiency; these results come at a time when they are particularly pertinent. Potential benefits of the PMCs structure for the automotive industries are as follows:

- Weight reduction, which may be translated into improved fuel economy and performance.
- Improved overall vehicle quality and consistency in manufacturing.
- Corrosion resistance.
- Part consolidation resulting in lower vehicle and manufacturing costs.
- Improved ride performance (reduced noise, vibration, and harshness).
- Vehicle style differentiation with acceptable cost.
- Lower cost of vehicle ownership.
- Lower investment costs for plants, facilities, and tooling-depends on cost/volume relationships.

However, there are areas where major uncertainties exist that will require extensive research and development prior to resolution. For example:

- High-speed, high-quality manufacturing processes with acceptable economics.
- Repairability.
- Satisfaction of all functional requirements.
- Particularly crash integrity and long-term durability.
- Recyclability and customer acceptance.

Fibers Used for Reinforcement of Polymer Matrix for Automotive

Carbon Fibers (CF)

Carbon fibers, also known as graphite fibers, can be based on three chemical sources: poly acrylo nitrile (PAN method is led by Japanese manufacturers), rayon (from the Indian manufacturer Grasim) and petroleum pitch. About 90% of the carbon fibers are produced or made from PAN. The exact composition of each precursor varies according to the recipe of manufacturers. The PAN based fibers give excellent mechanical properties for structural applications and the pitch fibers present higher modulus values and favorable coefficients of thermal expansion. The main benefits of carbon fibers are their low density and its high strength-to-weight ratio and stiffness. Its cost and brittleness is the main downsides as compared to the other fibers. At 46,500 tonnes of global annual demand for CF in 2013 has seen a growth of 15.1% CAGR since 2009 (26,500t). Global demand of CF is estimated to continue to grow to reach 89,000t by 2020. Regarding production capacities, the most important regions are North America (30%), Europe (24%) and Japan (20%). The top five CF manufacturers according to production volumes in 2013 were Toray, Toho, MRC, and SGL in that order. The composites design, analysis, and manufacturing tools will help for a reduction in the engineering cycle time and costs. Also, improve the quality while maintaining repeatability of parts being manufactured. The decrease in the projected cost of the carbon fibers will be mainly driven by:

- The development of a less expensive precursor material to produce the CF (textile PAN precursors or even lignin-based precursors instead of oil/gas based precursors to decouple the material price from oil price developments), results in 30% to 50% cost decrease for the raw material.
- A reduction in the processing cost for pre and part forming of 60% to 80% due to radical reductions in cycle times, for example; the development of fast curing resins and the resulting reduction of investment and labor costs per part.

Carbon fiber is the highest weight reduction potential (50% lighter than steel) but also has the highest cost (570% the cost of steel today). The effect of carbon fibers percentage on the weight of medium size car is shown in figure. The major techniques for cost reduction and their areas are shown in figure.

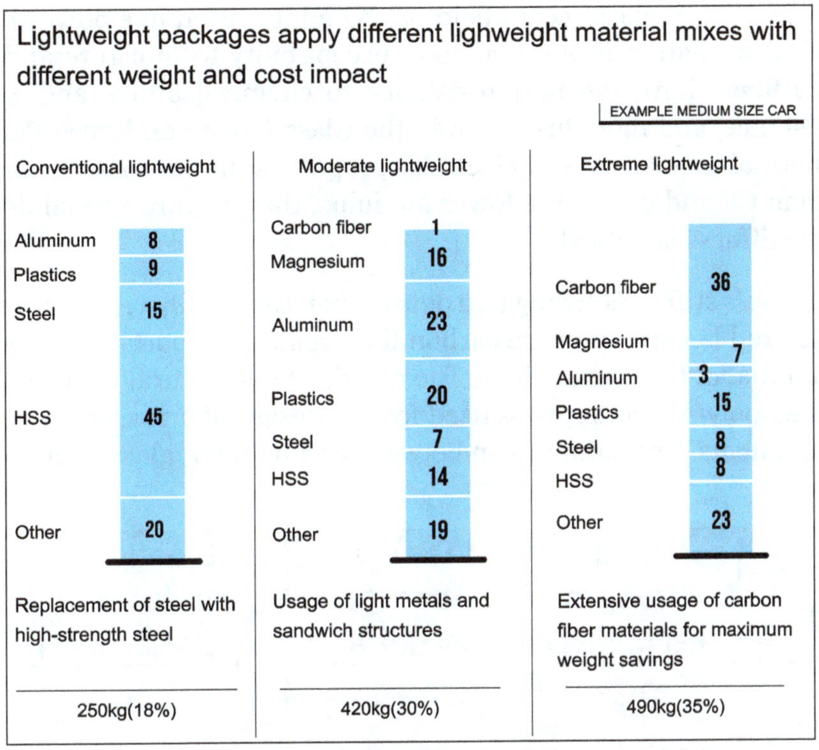

Weight saving in percentage by using lighter materials.

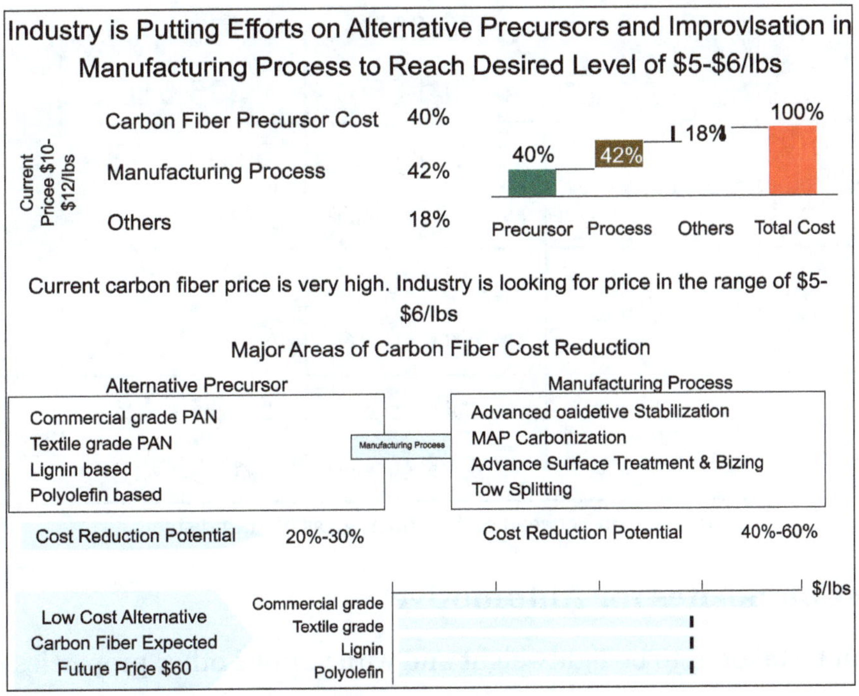

Reduction of cost of carbon fiber.

Glass Fibers (GF)

Glass fiber is based on an alumina-lime-borosilicate composition. The recipe can be varied resulting in different commercial compositions: E-glass (electrical), C-glass (chemical), R-glass, S-glass, and T-glass. In GF market, E-glass accounts for 90% and it is used in a polyester matrix and providing high electrical insulation properties, low susceptibility to moisture and high mechanical properties. C-glass fibers have the best resistance to chemical attack and S-glass has higher strength, heat resistance, and modulus. Overall, the Glass Fiber Reinforced Polymer (GFRP) exhibit very good thermal insulation and electrical properties and are also transparent; however, they are heavier than CF and due to the lower modulus, they require special design treatment in applications where stiffness is critical.

Because glass fiber is less stiff, less strong and denser than carbon fiber, glass fiber reinforced parts are typically thicker and heavier than the carbon fiber reinforced parts. The price of glass fiber is 5-10 times less than that of the carbon fiber. The most suitable natural fiber to be hybridized with glass fiber reinforced polymer composites used for the design of a passenger vehicle center lever parking brake components. The comparison between carbon fiber, glass fiber and steel are shown in figure.

Reinforcing Material	**Yield Strength** ksi (MPa)	**Tensile Strength** ksi (MPa)	**Elastic Modulus** ksi (GPa)	**Ultimate Strain** percent
Steel	40-75 (280-520)	60-100 (420-690)	29,000 (200)	10-15
Glass FRP	N/A	70-230 (480-1,600)	5,100-7,400 (35-51)	1.2-3.1
Basalt FRP	N/A	150-240 (1,035-1,650)	6,500-8,500 (45-59)	1.6-3.0
Aramid FRP	N/A	250-368 (1,720-2,540)	6,000-18,000 (41-125)	1.9-4.4
Carbon FRP	N/A	250-585 (1,720-3,690)	15,900-84,000 (120-580)	0.5-1.9

Comparison between carbon fiber, glass fiber and steel.

Metal Matrix Composites for Automotive

The materials can be tailored to be lightweight and with various other properties including are as follows:

- High specific strength and specific stiffness.
- High hardness and high wear resistance.
- Low coefficient of thermal expansion and High thermal conductivity.

- Low coefficients of friction.
- High energy absorption and a damping capacity.

In addition to these properties, new MMCs are being developed which have self-healing, self-cleaning and self-lubricating properties, which can be used to enhance energy efficiency and reliability of automotive systems and their components. The MMC system is generally designated simply by the metal alloy designation of the matrix then the material type followed by weight or volume fraction and form of the ceramic reinforcement. For example, 6061Al/30 vol. % SiC-p designates a discontinuously reinforced 6061 aluminum alloy with 30 volume % fraction of silicon carbide particulate. MMCs differ from the polymer or ceramic matrix composites in several ways. Some of these general distinctions are as follows:

- The matrix phase of an MMC is either a pure metal or alloy as compared to a polymer or ceramic.
- MMCs have higher ductility and toughness than the ceramics or ceramic matrix composites (CMCs) and also their respective unreinforced matrix.
- The reinforcement in the MMCs is increased the strength and modulus as is the case with PMCs. Reinforcement in CMCs generally provides improved damage tolerance.
- The MMCs have a temperature capability generally higher than the polymers and PMCs but less than the ceramics and CMCs.

Al or Al Alloy Metal Matrix Composite

Aluminium based metal matrix composites (MMCs) are "engineered materials" for automotive. This composite is composed by the addition of non-metallic (generally ceramic) particles and/or fibers in 5-50% by weight fraction to the metal. The additions non-metallic material can produce a considerable alteration in the mechanical and tribological properties of the base alloy. Tensile strength, yield strength and in some case fatigue strength can be significantly improved over the entire range of temperature. The Al MMCs also have enhanced physical property characteristics i.e. higher modulus, lower coefficient of thermal expansion, improved tribological characteristics and higher hardness versus unreinforced aluminum. Over the past 10 years, low-cost particulate reinforcements i.e. SiC, Al_2O_3, fly ash and graphite have been developed to reduce the cost of MMC materials. Also, there has been significant progress in the development of low-cost processing technique.

Magnesium Alloy Metal Matrix Composite

Magnesium is one of the lightest structural metals. The major advantage of Mg is its ability to reduce vehicle weight and enhance the performance of the vehicle. Magnesium parts can be tuned to those critical frequencies where noise, vibration, and harshness are reduced. At room temperature, Mg alloys have very low formability, but today's interest in Mg alloys for automotive applications is based on the combination of high strength and low-density property and for this reason, Mg alloys are very attractive as a structural material in all applications where weight savings are the main requirement. In automotive applications, the reduction in weight will improve the performance of a vehicle by reducing the rolling resistance and energy of acceleration, thus reducing the fuel consumption and moreover a reduction of the greenhouse gas CO_2 can also be achieved.

Parts of Automotive Fabricated by Composite Materials

Engine Block Cylinder Liners

In modern aluminium engine blocks, grey cast iron cylinder liners are used to protect the block. These weight about 9 kg per block but MMC liners weight 3.5 kg less. Al MMC liners can improve engine operating efficiency by reducing knock (heat transfer from the cylinder to water jacket is improved due to increased thermal conductivity). Blocks with MMC liners have increased rigidity as compared to the blocks with coated bores, which may be translated into increased cylinder roundness and reduced engine friction. Honda has successfully used in situ formation of an MMC cylinder via die-casting non-metallic cylinder preforms. In a new approach to achieve a further weight reduction, researchers will developed a new aluminium engine block which has the cylinder bore surface reinforced with short hybrid fibers of alumina and carbon.

Main Bearings

Copper-lead bearings are used in crankshaft main bearing caps can be replaced with lead free aluminium or copper matrix composites containing graphite particulates. Gr is nontoxic and the use of Gr reinforced Al or Cu composite bearings as a replacement for leaded copper reduces weight. The Gr reinforced MMC bearings also improve the wear characteristics because deformation of the Gr particulates results in the formation of a continuous graphite film, which provides self-lubrication of the component, allowing for improved component longevity.

Connecting Rods

With the advent of nanostructured materials, new materials have been developed with exceptional properties exceeding those expected for monolithic alloys or composites containing micron-scale reinforcements. For example, carbon nanotubes reinforced composites have ultrahigh strength and modulus. In another example reinforcements of only 10 vol. % of 50nm alumina (Al_2O_3) particulates to an Al alloy matrix using the powder metallurgy process increased yield strength to 515 MPa. This is 15 times stronger than the base alloy, 6 times stronger than the base alloy containing 46 vol. % of 29μm Al_2O_3 and over 1.5 times stronger than AISI 304 stainless steel.

Accessories

For components not exposed to extreme loading, further cost and weight reductions can be realized by reinforcing of fly ash (a waste by-product of coal power plants) in metal (e.g., aluminum, magnesium, lead, and zinc) matrix. Replacing components such as A/C pump brackets, alternator housings, timing belt/chain covers, valve covers, transmission housing and intake manifolds with Al reinforced with fly ash composites can reduce the vehicle cost and weight. Fly ash reinforced Al MMCs used for accessories can reduce emissions and save energy. It also reduces its coefficient of thermal expansion and increases its wear resistance along with making lighter and less expensive material.

Chassis

The performance of vehicle can affect by the strength and toughness of the chassis. Hollow ceramic microspheres reinforced metal matrix results in a syntactic foam product. Its density is about one half as compared to the matrix and it can able to absorb large amounts of energy per unit weight upon impact as compared to the monolithic alloys and open cell foams. Al-fly ash chemosphere syntactic foams being developed at UWM can be used in crumple zones to both increase torsional rigidity and increased energy absorption upon vehicle impact. In advanced automotive vehicles, syntactic foam also serves as a core material to increase the rigidity of thin gage sheet metal sandwich structure. High performance material like Kevlar honeycomb core material is cost prohibitive.

Bumper

Bumper is the one of the main part of an automotive vehicle having slightly more weight. We can employ composite materials in the bumper without sacrificing the safety. Polymers composite have been a major part of the automotive industries for several decades but the technical barriers and economic have constraint their use. At present, carbon fiber reinforced and glass fiber reinforced polymer composite are commonly used for automotive bumpers.

Leaf Spring

The Corvette was fitted with glass fiber reinforced epoxy polymer composite leaf springs, whose fatigue life is more than five times that of steel. Leaf springs developed by composite, give a smoother ride than steel leaf springs and also give more rapid response to stresses caused by road shock. Moreover, composite leaf springs have less chance of catastrophic failure and excellent corrosion resistance property. Another component i.e. brake drum, disc brake, bearing etc. can also be developed by the metal matrix composite which has good mechanical and tribological properties with lower weight.

Manufacturing and Assembly Cost Estimation

The cost estimation models must consider both the part manufacturing and assembly line. The total part cost (C_{tot}), including both manufacturing and assembly lines, is described as follows:

$$C_{tot} = C_{invest} + C_{tot-material} + C_{tool} + C_{running} \in \text{Assembly, manufacturing}$$

$$C_{invest} = f\,(\text{machines, presses, robots})$$

$$C_{tot-material} = f\,(\text{material utilisation and scrap})$$

$$C_{tool} = f\,(\text{tool cost})$$

$$C_{running} = f\,(\text{electricity, service, facilities, labor, etc}).$$

The initial material and process selection phase aims to estimate the cost of part manufacturing for a structure with low complexity and a set weight (normally comparison is done comparing the cost of a 1 kg structure). For the analysis of integral and differential design solutions, more detailed cost estimation is required by taking into account geometry, size, complexity etc. A manufacturing line is designed for each process considered in the above equation. Based on the operational time ($t_{operation}$) for each station, the annual work time (t_{annual}) and the annual planned volume (V), the specific number of annual operations available can be assessed for each station (n_{annual}) by:

$$n_{annual} = \frac{t_{annual}}{t_{operation} \times V}$$

The annual volume planned (V) for the production governs the number of robots and machines required in the process and for each station. In the cost models, a parallel line design must be considered due to the slow cycle times for composites and the often high annual volumes demanded by the automotive industry.

Aerospace Applications

Use of Composites in Aerospace Structure

It is to be realized that in order to meet the demands in table, it is necessary to have materials with a peculiar property-set. The use of composites has been motivated largely by such considerations. The composites-in particular, the advanced fibre reinforced composites using carbon or aramid fibres in polymer matrices offer several of these features as given below:

- Light-weight due to high specific strength and stiffness.
- Fatigue-resistance and corrosion resistance.
- Capability for high-degree of optimization: tailoring the directional strength and stiffness.
- Capability to mould large complex shapes in small cycle time reducing part count and assembly times. Good for thin-walled or generously curved construction.
- Capability to maintain dimensional and alignment stability in space environment.
- Possibility of low dielectric loss in radar transparency.
- Possibility of achieving low radar cross section.

These composite materials also have some inherent weaknesses:

- Laminated structure with weak interfaces: poor resistance to out-of-plane tensile loads.
- Susceptibility to impact-damage and strong possibility of internal damage going unnoticed.
- Moisture absorption and consequent degradation of high temperature performance.
- Multiplicity of possible manufacturing defects and variability in material properties.

Table: Features of aircraft structure.

Requirement	Applicability	Effect
Light-weight	All Aerospace Programmes	• Semi-monocoque construction: Thin-walled-box or stiffened structures. • Use of low density materials: Wood, Al-alloys, Composites. • High strength/weight, High stiffness/ weight.
High reliability	All Aerospace Programmes	• Strict quality control. • Extensive testing for reliable data. • Certification: Proof of design.
Passenger safety	Passenger vehicles	• Use of fire retardant materials. • Extensive testing: Crashworthiness.
Durability-Fatigue and corrosion Degradation: Vacuum Radiation Thermal	Aircraft Spacecraft	• Extensive fatigue analysis/testing. • *Al-alloys do not have a fatigue limit. • Corrosion prevention schemes. • Issues of damage and safe-life, life extension. • Extensive testing for required environment. • Thin materials with high integrity.
Aerodynamic performance	Aircraft Spacecraft	• Highly complex loading. • Thin flexible wings and control, surfaces. • *Deformed shape-Aeroelasticity. • * Dynamics. • Complex contoured shapes. • *Manufacturability: N/C Machining; Moulding.
Multi-role or functionality	All Aerospace Programmes	• Efficient design. • Use: composites with functional properties.
Fly-by-wire	Aircrafts, mostly for fighters but also some in passenger a/c	• Structure-Control Interactions. • * Aero-servo-elasticity. • Extensive use of computers and electronics. • * EMI shielding.
Stealth	Specific Military Aerospace Appl	• Specific Surface and Shape of aircraft * Stealth coatings.
All-weather operation	Aircraft	• Lightning protection, erosion resistance.

Even after accepting these weaknesses, the projected benefits are significant and almost all aerospace programmes use significant amount of composites as highlighted in figure. All this is, of course, not without its share of hassles. Challenges of using composites on such a large scale are many. The composites are not only new but also non-conventional: they are anisotropic, inhomogeneous, have different fabrication and working methods and also different controls for quality assurance. They have a complex material behaviour under load requiring new and complicated analysis tools. Moreover, the behaviour is not always predictable by analysis and this makes reliance on several expensive and time consuming tests unavoidable.

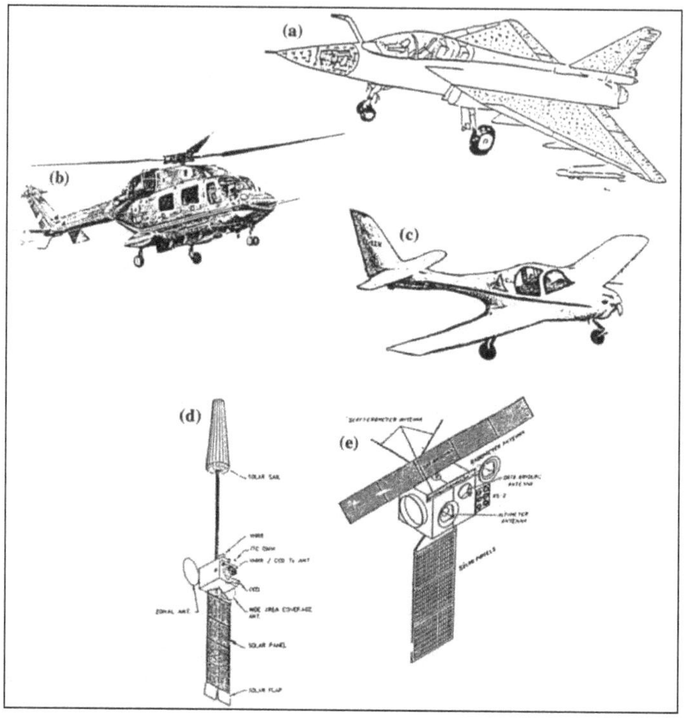

Figure shows the use of composites in various aerospace programmes: (a) A military aircraft with wings, fin, control surfaces, and fuselage panels made in carbon-epoxy, radome in aramidpolyester. (b) An advanced helicopter using aramid, glass and carbon composites in all structural parts. (c) An all composite 2-seater transport aircraft using glass-epoxy supplemented with carbon. (d) And (e) satellites using carbon composites for solar panels, antennae, etc.

The routes to meet these challenges have evolved around use of the advances in computer technology and analysis methods to implement schemes based on computer aided design, computer aided engineering, finite element methods of analysis and building computer interfaces amongst all aspects of development, namely, design, analysis and manufacturing. These should provide fast transfer of information including graphics and accurate analysis methods for a reasonable prediction of complex behavioural patterns of composites. It is only by harnessing the vast computational power for various purposes that the aircraft structural design of today can meet the challenges posed by the required performance.

Materials for Aerospace Composites

The material systems which have been considered useful in aerospace sector are based on reinforcing fibres and matrix resins given in tables, respectively. Most aerospace composites use prepregs as raw materials with autoclave moulding as a popular fabrication process. Filament winding is popular with shell like components such as rocket motor casings for launch vehicles and missiles. Oven curing or room temperature curing is used mostly with glass fibre composites used in low speed small aircraft. It is common to use composite tooling where production rates are small or moderate; however, where large number of components is required, metallic conventional tooling is preferred. Resin injection moulding also finds use in special components such as radomes. Some of the popular systems are given in table along with the types of components where they are used in a typical high-performance aircraft.

Concerns with Composite Usage

The concern in use of composites arises mainly due to demands of high degree of reliability and safety of aerospace structures as against the complexity of composite behaviour and consequent difficulties in building prediction models. This creates an excessive reliance on testing at all stages: design and development, proving and certification, and in-service inspection and repairs. The costs of such testing are sometimes enormous and this had led to some scepticism in use of composites.

Impact Damage and Damage Tolerance

The laminated structure of the composites and the fibre-matrix interfaces provide weak interfaces for delamination or de-bonding to take place. This is further aggravated by practical structural features such as discontinuous plies to create thickness changes and sharp bends required in stiffening members. Of particular concern is the proneness exhibited for damage due to impact. The issue is not merely the reduction in strength (particularly in compression) but also that the damage is inside the material and not visible at the structure. This is particularly so where the impact is due to blunt objects at low to medium velocities. Common instances are dropping of tools, hail-stones, runway debris and impacts and jolts while handling (even before the assembly of the aircraft). Such hidden damage can be extensive both in terms of planar dimensions and through the thickness. The damage mostly occurs as delamination, but may sometimes be accompanied by fibre-breaks in back plies which are not visible from outside. In the shop, such damages can be found by ultrasonic C-scan method and 'barely visible impact damage' can cause a reduction in compression strength by almost 60%. The fatigue resistance of carbon composites stands it in good stead, however, and no further significant reduction in strength or growth of damage is observed under in-plane loads.

Table: Reinforcing fibres commonly used in aerospace applications.

Fibre	Density (g/cc)	Modulus (GPa)	Strength (GPa)	Application areas
Glass				
E-glass	2.55	65-75	2.2-2-6	Small passenger a/c parts, aircraft interiors, secondary parts; Radomes; rocket motor casings.
S-glass	2.47	85-95	4.4--4	Highly loaded parts in small passenger a/c.
Aramid				
Low modulus	1.44	80-85	2.7-2-8	Fairings; non-load bearing parts.
Intermediate modulus	1.44	120-128	2.7-2-8	Radomes, some structural parts; rocket motor casings.
High modulus	1.48	160-170	2.3-2.4	Highly loaded parts.

Carbon				
Standard modulus (high strength)	1.77-1.81	270-300	3.0-3-5	Widely used for almost all types of parts in a/c, satellites, antenna dishes, missiles, etc.
Intermediate modulus	1.77-1-80	270-300	5.4-5.7	Primary structural parts in high performance fighters.
High modulus	1.77-1-80	390--45	2-8-3-0 4-0-4-5	Space structures, control surfaces in a/c.
Ultra-high strength	1 - 8 0 - 1.82	290-310	7-0-7-5	Primary structural parts in high performance fighters, spacecraft.

The current philosophy to handle impact damage problem is as follows: (i) Design the structure to have alternate load paths to have damage tolerance against impact of moderate severity. This is generally taken care by designing the structure as a framework of stiffening members or as boxes; (ii) Lower the design allowable strength values to an extent where the 'barely visible impact damage (BVID)' can be sustained even at the highest load and for all the time with no degradation in performance; (iii) Any damage that exceeds the BVID level (i.e. visible damage) may lower the immediate performance and should be repaired immediately. The basic safety of aircraft with damage is ensured due to (i) and (iv) the structure may not cater to very severe impact.

There is, of course, a penalty in lowering the allowables but for the present systems, this is considered to be not too excessive in view of the similar reduction of allowables required for fastener holes. With improved processing to get large parts integral with stiffeners and other complex shapes and with availability of high strength fibres the limitations due to impact damage would be more perceptible, and prohibitive. Another consequence of the impact damage issue which the aeronautical community is, perhaps, not yet fully exposed to is in terms of setting up of the inspection intervals and defining levels of repairs etc. when the presently developed aircraft go in full service. Extensive studies and gathering of experience through testing is presently underway to tackle this problem.

Table: Polymeric matrices commonly used in aerospace sector.

Thermosets				Thermoplastics
Forms cross-linked networks in polymerization curing by heating				No chemical change
Epoxies	Phenolics	Polyester	Polyimides	PPS, PEEK
• Most popular. • 80% of total composite usage. • Moderately high temp. • Comparatively expensive.	• Cheaper. • Lower viscosity. • Easy to use. • High temp usage. • Difficult to get good quality composites.	• Cheap. • Easy to use. • Popular for general applications at room temp.	• High temp • Application 300°C. • Difficult to process • Brittle.	• Good damage • Tolerance. • Difficult to process as high temp 300-400 °C is required.
• Low shrinkage (2-3%). • No release of volatile during curing.	• More shrinkage. • Release of volatile • during curing.	• High shrinkage (7-8%).		

• Can be polymerized in several ways giving • varieties of structures, • morphology and wide range of properties.	• Inherent stability for thermal oxidation. • Good fire and flame retardance. • Brittle than epoxies.	• Good chemical • Resistance. • Wide range of properties but lower than epoxies. • Brittle. • Low Tg.		
• Good storage stability to make prepregs.	• Less storage stability difficult to prepreg.	• Difficult to prepreg.		• Infinite storage life. But difficult to prepreg.
• Absolute moisture • (5-6%) causing swelling and degradation of high temp properties. • Also ultra violet degradation in long term.	• Absorbs moisture but no significant effect of moisture in working service range.	• Less sensitive to moisture than epoxies.		• No moisture absorption.
• Density (g/cm3) 1.1-1.4.	• Density (g/cm3) 1.2-1.4.	• Density (g/cm3) • 1.1-1.4.		• Density (g/cm3) 1.3-1.4.
• Tensile modulus 2.7-5.5 GPa.	• Tensile modulus 2.7-4.1 GPa.	• Tensile modulus • 1-3-4.1 GPa.		• Tensile modulus • 3.5-4.4 GPa.
• Tensile strength • 40-85 MPa.	• Tensile strength • 35-60 MPa.	• Tensile strength • 40-85 MPa.		• Tensile strength • 100 MPa.

There is also a problem of bullet impact on composite components especially those carrying aircraft fuel. Apart from the damage to composite, there is an apprehension that the pressure wave due to such an impact will cause a hydraulic ram effect in the fuel which will compress the air-fuel mixture in the space in the tank and lead to explosion.

Environmental Degradation

The presently used epoxy resins absorb about 5-6% moisture by weight when fully saturated. This leads to about 1.5-1.8% moisture weight gain in carbon-epoxy composites with the usual 60% fibre volume fraction. In practice, under the normal operating conditions, the maximum equilibrium moisture gain in an aircraft component can be about 1.0-1.4%. This moisture gain can cause (a) swelling and dimensional changes, (b) lowering of the glass transition temperature (T) of the resin matrix, and (c) degradation of matrix dominated properties of composites such as shear and compression strengths.

The dimensional changes and weight gain by itself are generally not significant in many aircraft structures but may be of considerable significance where extreme precision is required such as in antennae panels and in some satellite structures. The major concern in aircraft structures is the degradation of the shear and compressive strength properties--particularly at high temperatures close to T which in itself is now reduced due to moisture absorption. The design of a structural component, therefore, generally proceeds by reducing allowables for moisture degradation. This in itself is rather straightforward. More complicated problems, however, arise in getting the proof of design and the certification of airworthiness. It is generally prohibitively expensive and often impractical to test actual components in the environmentally worst conditions to demonstrate adequate strength under those conditions and recourse has to be taken to extrapolation from data obtained from coupons and smaller structures. Such extrapolations are not always straightforward and extensive test programmes are required to demonstrate proof of design. A typical approach taken is the 'building-block approach' of generating data on various levels of increasing complexity. Table illustrates a typical test programme for a major aircraft component like a wing. Note the decrease in number of specimens as the complexity of the specimen geometry increases.

This single issue of environmental degradation due to moisture absorption has made development of composite components for aerospace quite expensive and tedious. Moreover, associated with the already complex behaviour of composites, the environmental degradation issue leaves some lingering doubts about behaviour of composites, particularly in long term. Apart from the moisture absorption, the other significant aspects relate to the UV degradation and radiation effects in long term. These are particularly important for space structures. The current studies on the subject have provided some solutions to these problems even though the concern about long term behaviour exists.

Table: Typical composite material systems in aerospace.

Material system	Application area
• 175 °C curing high strength-carbon-epoxy -Zero-bleed (neat resin content) UD prepregs • 5HS or 8HS bi-directional fabric prepreg -toughness, good out-life and shelf-life	Structural components of fighter aircraft and helicopters, e.g. wing skins, spars, fin, rudder, elevons, doors, etc.
• 175 °C curing intermediate modulus carbon with epoxy + BMI/cynate-ester • Zero-bleed (neat resin content) UD prepregs • 5HS or 8HS bi-directional fabric prepreg • High toughness, good out-life and shelf-life • Low environmental degradation	Frames, stiffeners, rotor blades.
• 120 °C curing HS-carbon-epoxy • Zero-bleed (neat resin content) UD prepregs • 5HS or 8HS hi-directional fabric prepreg • Toughness, good out-life and shelf-life	Structural components of helicopters or transport aircraft, e.g. spars, fin, rudder, elevons, doors, etc. Frames, stiffeners.
• Aramid fibre in low-loss polyester/cynate esters	Radome.
• Cu-mesh epoxy prepreg	For Lightning Strike protection Wing-skin, others.
• E-glass fabric in epoxy resins • High temp curing • RT/moderate temp curing	Fighters fairings, fin-radome, drop-tanks. Small transport aircraft structhral components: Fuselage, wing, other.

Advances in Materials for Composites

Reinforcements

The carbon fibre technology continues to improve harnessing the versatility of carbon fibre and new varieties in terms of better combinations of modulus and strength are becoming available.

The developments seem to be in two directions: one, for aircraft applications, is aimed basically at higher strength (> 5 GPa) with concurrent improvements in modulus to a moderate level (> 300 GPa) and the other, for space applications, is aimed at high modulus (>500GPa) with moderate strength (-3.5 GPa). The higher failure strain for the fibre is expected to result in composites with better damage tolerance. The developments in aramid fibres also aim at higher modulus with concurrent increase in strength. However, the major thrust in improving reinforcements for composites comes from the requirements of multidirectional weaving. Several processes (weaving, knitting, braiding) have been developed for this purpose and preforms with multi directionally woven fibres have now been made. Simplification and cost reductions appear to be the major motives for further developments.

The higher properties of basic fibres (such as carbon) cannot, however, be fully exploited in the composite without concurrent developments in the matrix materials and the intermediate products such as prepregs or preforms. It is to be noted here that the carbon fibre composites which use a carbon fibre with a strength of 3 GPa as reinforcement result in an allowable stress of only 0.3 GPa in a composite. Significant scope thus exists for translating high fibre properties into high performance of composites.

Matrix Resins

A significant effort in improving composites is focused on improving matrix materials. The two major concerns mentioned earlier viz. impact damage tolerance and hygrothermal degradation, provide the main motivation for improvement. A major direction of improvement appears to be an improvement in the toughness which should result in higher resistance to delamination and against impact. High failure strain of matrix resin would help in translating the higher performance of the improved fibre to the composite. Higher resin shear modulus would help in achieving better transfer of load from fibre to resin and again to fibre and should therefore improve compression strength.

For polymeric materials a possible figure of -5 GPa should be achievable as against the current resins with shear modulus of about 2 GPa. As far as hygrothermal degradation is considered, newer systems based on cynate ester look very promising and some of these have already found some application. Another route being investigated is the use of thermoplastic resins and their blends. Poly-ether-ether-ketone (PEEK) has been considered very promising, but the industry needs to resolve the problems associated with high temperature (> 350 °C) processing of the material. Current approaches to new resins appear to be directed towards producing polymeric systems which can be processed in the way composites industry is used to (such as autoclave curing up to 180 °C). Two promising classes of materials under development are given below:

- Polymerizable liquid crystalline monomers: These should result in thermoset resins with high fracture toughness and $T_g \approx 170°C$ with high retention (= 90%) under hot-wet conditions. Comparing to thermoplastic PEEK, such resins will have almost similar fracture toughness (composite Glc = 800 J/m2) with the advantage of conventional processing. The system is being looked into for potential applications in civil transport aircraft. The approach for development of these new resins that is being followed is first, to synthesise controlled molecular weight backbones consisting of aromatic ethers, aromatic esters or rigid aliculic systems with hydroxy end groups and then to end-cap these backbones with reactive end groups like cynate ester group or glycidyl ethers.

- Phthalonitrile resins for high temperature applications which can be cured in conventional manner (at 180-200 °C) but can be post-cured, albeit in inert atmosphere, at high temperatures up to 600 °C. Compared with the presently available PMR resins for high temperature (250-350 °C) applications, the new resins will have better processibility, good fracture resistance, better strength and modulus and very low moisture absorption.

Table: A typical approach for design validation (example: composite wing).

		Test specimens		
Level of testing	Types of tests	Total	(Hot/W)	Remarks
1. Coupons	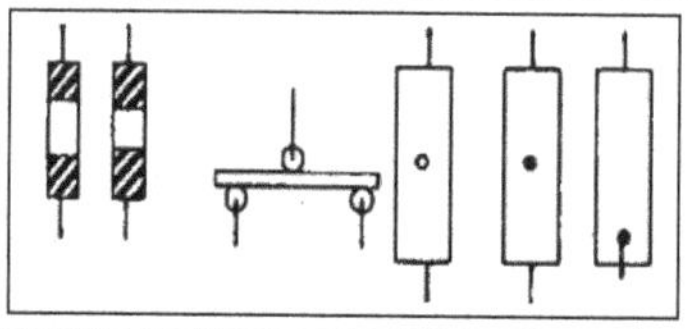	1200	(600)	Basic material properties (tension, compression, shear, bearing, etc.) at RTD and HTW. Mean and B-values.
2. Structural features	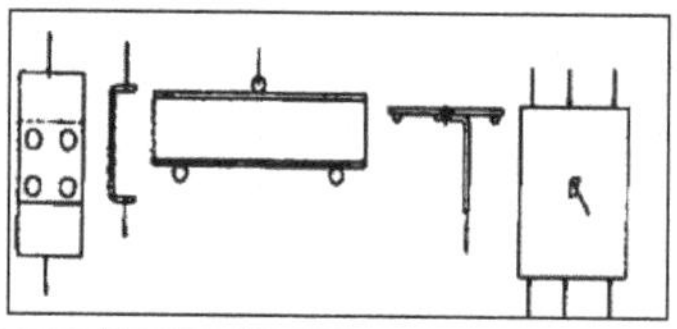	500	(180)	Estimation of practically realizable strengths, variability.
3. Structural test boxes	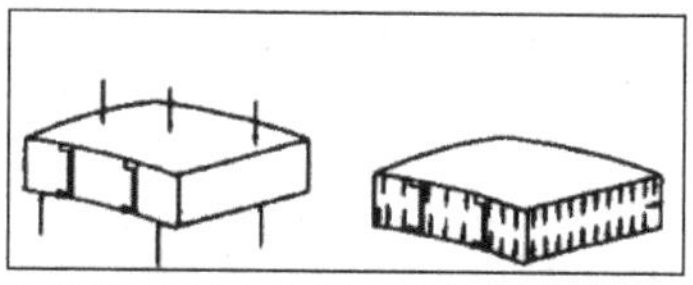	17	(4)	Simulation of structure and loading.
4. Full-scale component (Half-wing)	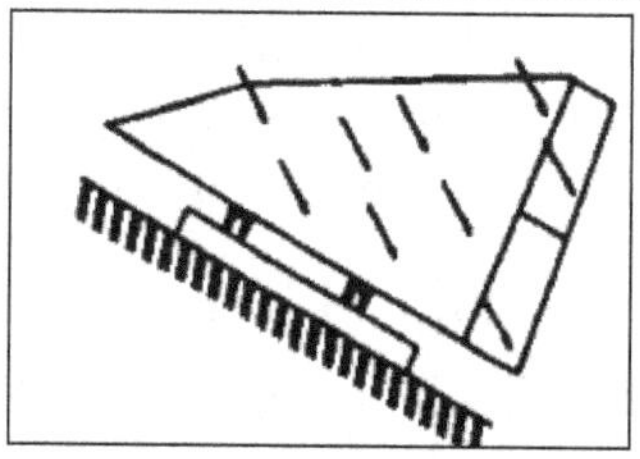	1	(0)	Actual structure and loading.

Advantages of Composites in Aviation Industry

Composite materials are used more and more for primary structures in commercial, industrial, aerospace, marine and recreational structures. Advanced composites do not corrode like metals – the combination of corrosion and fatigue cracking is a significant problem for aluminium commercial fuselage structure. Composites today have a wide array of benefits in the aerospace and defence industry. Resulting fuel efficiency gained by an aircraft is becoming increasingly important with today's soaring fuel prices. Other positive attributes include excellent fatigue and corrosion resistance and good impact resistance. Composite usage has increased across aerospace and defence verticals, and some segments are expected to grow significantly in the next 20 years. There are some advantages of composite materials:

- Weight reduction: Savings in the range 20% - 50% are often quoted.
- Mechanical properties can be tailored by 'lay-up' design, with tapering thicknesses of reinforcing cloth and cloth orientation.

- High impact resistance: Kevlar (aramid) armor shields planes, too, for example, reducing accidental damage to the engine pylons which carry engine controls and fuel lines.
- High damage tolerance improves accident survivability.
- 'Galvanic' electrical: Corrosion problems which would occur when two dissimilar metals are in contact (particularly in humid marine environments) are avoided. Here non-conductive fibreglass plays a roll.

Disadvantages of Composites in Aviation Industry

- Some higher recurring costs.
- Higher nonrecurring costs.
- Higher material costs.
- Non-visible impact damage.
- Repairs are different than those to metal structure.
- Isolation needed to prevent adjacent aluminium part galvanic corrosion.

Composite materials maximise weight reduction as they typically are 20% lighter than aluminium and are known to be more reliable than other traditional metallic materials, leading to reduced aircraft maintenance costs, and a lower number of inspections during service.

Usage of Composite Materials in Aerospace Industry

Composite materials can provide a much better strength-to-weight ratio than metals: sometimes by as much as 20% better. The lower weight results in lower fuel consumption and emissions and, because plastic structures need fewer riveted joints, enhanced aerodynamic efficiencies and lower manufacturing costs. The aviation industry was, naturally, attracted by such benefits when composites first made an appearance, but it was the manufacturers of military aircraft who initially seized the opportunity to exploit their use to improve the speed and manoeuvrability of their products.

Weight is everything when it comes to heavier-than-air machines, and designers have striven continuously to improve lift to weight ratios since man first took to the air. Composites materials played a major part in weight reduction, and today there are 3 main types in use: carbon fibre, glass and aramid – reinforced epoxy. There are others, such as boron-reinforced (itself a composite formed on a tungsten core). Composites are versatile, used for both structural applications and components, in all aircraft and spacecraft, from hot air gondolas and gliders, to passenger airliners or fighter planes. The types have different mechanical properties and are used in different areas of aircraft construction.

Carbon fibre for example, has unique fatigue behaviour and is brittle, as Rolls Royce discovered in the 1960's when the innovative RB211 jet engine with carbon fibre compressor blades failed catastrophically due to bird strikes. In an experimental program, Boeing successfully used

1500 composite parts to replace metal components in a helicopter. The use of composite-based components in place of metal as part of maintenance cycles is growing rapidly in commercial and leisure aviation. Overall, carbon fibre is the most widely used composite fibre in aerospace applications.

Marine Applications

Watercraft, submersibles, offshore structures, and other marine structural components are exposed to relevant environmental challenges. Therefore, materials exhibiting elevated resistance and requiring few or no maintenance for extended periods of time are generally under consideration. Additionally, being lightweight and corrosion resistance are mandatory for vessels, to meet design requirements and perform with speed and unfailing reliability. Composite materials, with specific reference to fiber-reinforced polymers (FRP), have been extensively incorporated in the marine industry following their early applications after World War II when composites were designed to overcome corrosion issues experienced with steel, aluminum, and wood. Weight reduction was, and remains, a key aspect, particularly for topside weight found in commercial vessels. Over the years, composites have been developed to meet the growing and variable demand coming from the marine sector, still preserving a relevant usage for manufacturing of consolidated elements, including gratings, ducts, shafts, piping, and hull shells. To date, composites are employed in all areas of the marine sector and for a variety of components and structures, namely hulls, bearings, propellers, hatch covers, exhausts, topside structures, radomes, sonar domes, railings, vessels of all types, valves and other subsea structures. In recent years, racing powerboats made of composites are becoming more and more common for long-lasting high performance and safety.

Salt and corrosive seawater are especially damaging to watercraft and structures. Given the capability of composites to resist corrosion and fatigue, maintenance requirements can be significantly reduced by using thermoset composites in specific components, such as fixings and connectors. The usage of thermoset resins as matrix material is typically associated with:

- The strength-to-weight ratio for lighter-weight durability.
- Dimensional stability for lasting performance.
- Corrosion resistance for deterioration-free operation.
- Lighter weight to reduce operating costs and improve efficiency.
- Possibility to combine multiple parts into one and reduced maintenance.
- Sound baffling for a less noisy operating environment.
- Design flexibility for use in complex shapes.
- High levels of acoustic transparency for radomes and sonar domes.
- Enhanced stiffness for stability.

Application of Composite Materials for Hulls, Ships, and Submarines

The nautical industry has experienced an evident technological evolution and, today, the manufacturing of hulls is based on latest-generation production systems. The introduction of advanced composite materials represented a milestone innovation in the manufacturing of boats and vessels. Composite technology has allowed manufacturers to improve the quality of products obtaining stiff and light structures, with benefits in terms of sailing performances and working life. In addition, thanks to the possibility to automatize and speed up manufacturing procedures, the adoption of composite materials has lowered the overall costs making the end product affordable to a wider public. Stiffer hulls and decks are the main applications where the shipbuilding industry has adopted composite sandwich structures, composed by two skins with high stiffness and strength placed on the external faces of a component and by a soft and thick core. Stiff skins provide high bending stiffness; meanwhile, the core supports the shear and compressive stresses and stabilizes the skins preventing global and local instabilities.

The weight reduction results in larger cargo capacity, fuel-saving, lower inertia, and increased ship stability and buoyancy. In addition, FRPs show satisfactory corrosion resistance in the marine environment and require less maintenance. In marine sandwich structures, aramid, carbon and glass FRP materials are commonly used as skins instead of metals. Polymeric foams (i.e., polystyrene and polyvinyl chloride (PVC)) and honeycomb are mainly used as core material. The combination of aramid fibers (such as Kevlar) or carbon to reinforce high-quality resins and ultra-light cores (such as Nomex), results in boats with unmatched mechanical properties and dimensions. On the other hand, composite sandwich structures have a higher cost compared to conventional materials and the manufacturing processes are more labor-intensive, limiting their widespread use.

Naval Applications

Military Applications

The first application of glass-reinforced plastic (GRP) for naval shipbuilding is dated back to 1973 with the mine-countermeasures vessels (MCMVs) HMS Wilton, characterized by a length of 60 m and a full load of 725 tons. It was realized by the Vosper-Thornycroft company and mass-produced by the British Navy since 1975. The GRP ship proved to be able to withstand the extreme working conditions in the Ocean as well as impacts and fires.

The subsequent 'Sandown' class project of the United Kingdom Navy aimed to reduce both the weight for a greater weapon capacity and the costs, to improve the international competitiveness of the project. Oversized components were redesigned (acceptably reducing safety margins) still preserving acceptable reliability. Furthermore, bolts between frames and hull were removed. In 1975, an R&D program was jointly launched by France, Belgium, and the Netherlands to build 40 ships, providing a technology sharing by the involved countries. In particular, main machinery machines, auxiliary machines, and weapon equipment were realized by Dutch, Belgian and French companies, respectively. The DCN Diving Company provided the technologies for shipbuilding using GRP for hulls, decks and partitions, pins to join the hull and the reinforcements. Sandwich structures consisting of GRP skins with a PVC foam core were used in GRP MCMVs developed by Karlskrona shipyard in Sweden between 1970 and 1984.

The technical solution allowed in meeting the requirements of fatigue resistance, non-magnetic behavior, and economic and technical feasibility. Other examples are the production of catamaran hull for MCMV using sandwiches (foam-based core and GRP skin) by the Royal Australian Navy. The hull of the HMAS Rushcutter was realized using a 60 mm thick high-density PVC foam core and 8 mm thick GRP. More specifically, the skin panels were manufactured combining chopped strand mat and woven roving reinforcements, in order to maximize the bond strength, fire resistance, and inter-laminar shear strength. Figure reports the evolution over the years of the length of ships entirely produced using composite materials for the main three classes of vessels.

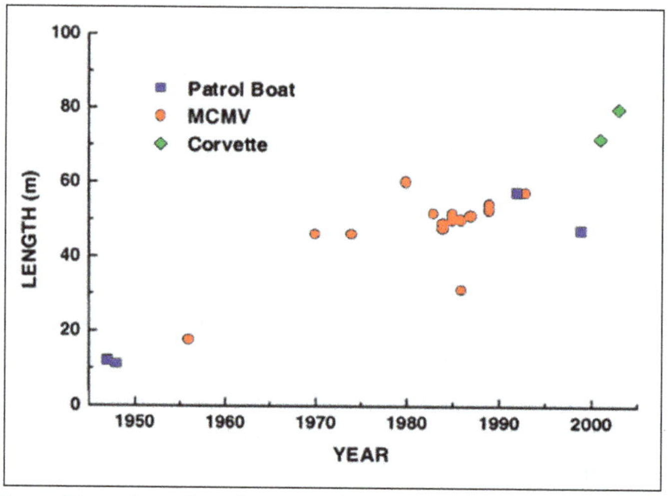

Figure: Plot of vessel length against the year of construction for all-composite patrol boats, mine-countermeasures vessels (MCMVs) and corvettes against the year of construction.

The corvettes are the longest ships currently built-in FRP. In this segment, the multipurpose Visby corvettes, designed and realized in the framework of the YS-2000 project by the Swedish Navy, is an excellent example of composite material application. With a length of 72 m, a width of 10.4 m and a full load of 620 tons, Visby was one of the largest ships in the fleet. It was designed for surveillance, combat, minesweeper, countermeasures and anti-submarine war operations. The Visby corvette was designed entirely in composite materials using sandwich panels with carbon and glass reinforcement and vinyl resin matrix. Indeed, composite materials provided robustness, resistance to underwater shock loads and low magnetic capacities combined with lightweight nature. Visby was the first ship with a significant content of carbon fiber composites within the hull. Although the cost of carbon fibers was at least five times higher than that of glass limiting their use in large naval structures, it was found that by using some carbon fibers in the composite, beside the adequate electromagnetic shielding, the weight of the hull was reduced by about 30%, without significantly increasing the manufacturing cost. The achieved weight-saving translated into a reduction in fuel consumption.

The Royal Singapore Navy and Sweden Kockums AB shipyards designed a new generation of corvette patrol, known as the New Generation Patrol Vessel (NGPV) class, made of FRP. The designed ship is 80 m long with a full load weight of 1016 tons. Technological developments in the design and manufacturing of composites have led to a steady increase of the lengths of the ship and currently, there are fully composite ships up to 80/90 m in length. Recent studies forecasted that hulls of medium-sized warships, entirely made of composite materials, could easily reach a length of 120/160 m in 2020. However, traditional materials (aluminum and steel) still remain the main choice for corvettes due to reduced manufacturing costs and higher reliability.

The United States Navy first introduced a fairwater entirely made of GRPs on the conning tower of a "Guppy" class submarine. These parts, traditionally made of aluminum, undergo severe corrosion and require continuous on-going maintenance and repairs. The composite ones resulted in more durable requiring reduced maintenance. As a result, more than 25 Guppy class submarines were equipped with composite structures between 1950 and 1970. Composite materials are also considered for internal applications within a pressurized submarine, including bulkheads, decks, hatches, main propulsion shaft, ballast tanks, tanks, machines, pumps, valves and piping. Figure summarizes the applications on military ships and submarines where the composite materials are currently employed or under development.

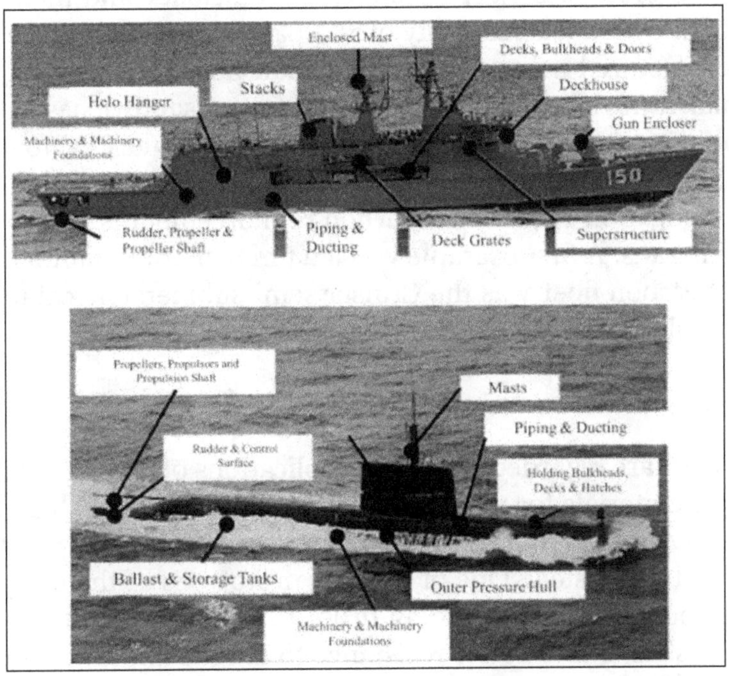

Military applications of fiber-reinforced polymers (FRP) structures to naval ships and submarines.

Civil Applications

Apart from combat ships, high-performance materials are now applied also in civil sectors. GRPs are worldwide used to produce vessels of fishing boats, small boats, hovercraft, and catamarans. Japan is by far the leading country in terms of the number of active GRP vessels. In 1984, Japanese GRP fishing boats represented 60% of the total registered vessels, including some hulls up to 45 m in length. Apart from Japan, large GRP fishing boats have been produced also in European and Australian shipyards.

The applications GRP for small boats include yachts, sailboats, barges, and lifeboats. GRP is the predominantly used material representing approximately 80% of the hulls, for ships long up to 20 m. GRP materials offer significant advantages in terms of reduced weight and production costs compared to the steel as well as minimal maintenance costs, thanks to the notable resistance to marine organisms and corrosion. Dutch "Mulder and Rijke" shipyards have developed a glass fibers/polyurethane composite lifeboat with elevated flame-retardant properties. In fire-resistance tests the boat was immersed in flames for 10 min: the outside temperature of 1500 °F (816 °C) was registered, meanwhile the internal temperature had reached only 15.5 °C.

Hovercraft ships represented the main application of fiber-reinforced plastics in the marine sector for a long time. For instance, the 15.5 m hull of the Hovermarine HM2 was manufactured in 1969 using sandwich structures made of fiber-reinforced composite laminates with PVC and polyester foam cores. The first version of 24 m Vosper Hovermarine's HM221 ferry, introduced in 1984, has been entirely built using sandwich structures with GRP skins and PVC foam core for the hull and GRP panels for the finishing. More recently, the Norwegian Brodrene produced the SES Norcat using the GRP sandwich reaching a length of 32.2 m, a gross tonnage of 290 tons and a maximum speed of 42 knots.

Catamarans are mainly used for fast ferries, offshore and oceanographic sailing ships and small pleasure boats due to its good passenger capacity, fuel efficiency and high reliability. The CIRR 27P CAT (27 m in length with a load capacity of 184 passengers), introduced in Norway in 1985, was the first ship manufactured using GRPs. Norwegian Boat service Verft produced a series of multi-purpose reconnaissance ships using aramid-reinforced sandwiches with sizes ranging from 23 to 38 m. The combination of aramid fiber laminates (skin) and PVC foam (core) provided an increase in hull tensile strength of the five times if compared to steel and in a reduction in noise, crucial for seismic investigation operations. Carbon reinforcement is also used in the production of catamarans and trimarans. A first example was a 24 m trimaran manufactured in 1976. A subsequent application of carbon fiber was the Cruiser semi-submersible catamaran built by Mitsui Engineering and Shipbuilding using hybrid carbon-glass reinforcement.

Boat Components

As far as boat components are regarded, the main applications of FRP are in masts and propellers. The usage of composite materials for the construction of the masts of ships started only in 1990, when a preliminary study showed that the usage of composite materials was the solution to several problems encountered using steel. One of the first prototypes of the composite mast was built with a hybrid glass/carbon fiber composite, in order to enhance both ballistic performance and rigidity. The composite shaft offers better resistance to fatigue and corrosion, providing also the vibration damping and air blast damage resistance. On the other hand, cost increased by 50% with respect to traditional ones. In 1995, the US Navy launched the Advanced Enclosed Mast/Sensor (AEM/S) program to develop the next generation of ship masts. The AEM/S system was sized 28 m in height high and 10.7 m in diameter with a hexagonal shape, resulting in the largest composite structure on a US navy ship at that time. The peculiar design, coupled with the hybrid nature of the reinforcement, improved the performance of antennas and sensors still maintaining an acceptable radar cross-section.

The Royal Navy and Vosper Thornycroft shipyard (UK) implemented the Integrated Technology Mast (ITM) using FRP in 1996. Similarly to the AEM/S project, the ITM consisted of a sandwich structure manufactured by redesigning the shape to reduce electromagnetic interference, with an additional 10/20% weight reduction compared to the steel masts. The ITM improved the stealth capability of the warship and provided a better environmental and electromagnetic shielding with respect to conventional steel antennas. ITM technology has been installed on ships since 2011. In Germany, hollow shafts are made by winding multiple layers of continuous glass and carbon filaments immersed in epoxy resin on wood/aluminum spindles.

Propellers in ships and submarines are usually made of nickel-aluminum-bronze (NAB) alloy, due to the outstanding corrosion resistance and high yield strength. On the other hand, the manufacturing of propeller blades in NAB alloy is expensive due to their complex shape and some issues related to the poor acoustic damping properties leading to excessive vibration noise are unsolved. These limitations pushed the naval engineers to evaluate the feasibility of alternative materials to NAB alloys to fabricate propeller blades, being stainless steel, titanium alloys and FRP the most notable candidates.

Composite propeller blades are usually made with glass or carbon fibers laminate, whereas a thin layer of polyurethane or NAB or stainless steel can be adopted to reduce impact damages to the blade. The FRP blades are generally glued or fastened to the metal hub of the propeller, although composite based hubs have also been realized. Early composite propellers, up to 2 m in diameter, were tested in Soviet fishing boats during the 1960s, and then they have been introduced in large commercial ships in the early 1970s reaching diameters of approximately 6 m. In the same years, prototype applications of such components were tested in hovercrafts. Composite propellers reported similar performances as metal ones in terms of navigation speed, fuel consumption, engine workload, and life cycle. Additionally, they reduced the amplitude of the resonance vibrations in the engine and transmission shaft by about 25%, with consequent reduction of hull vibrations and noise. On the other hand, composite propellers present higher manufacturing cost, large deviations of the blade tip and lower impact resistance than those made of NAB.

Apart from masts and propellers, some secondary elements, especially in the warships, are nowadays produced replacing metal with polymer composites. Some examples are rudders, bulkheads, decks and hatches, discharge funnels, protection systems and shields, engine components, and heat exchangers.

Materials for Naval Applications

Resins for Matrices

Boat builders require resins that have good thermal and mechanical properties and, at the same time, can be easily processed during the manufacturing of composite parts. Suitable resins should be easy to laminate, with room temperature curing, and compatibility with the reinforcing materials. Vinyl esters have mechanical properties and prices that are positioned halfway between polyesters and epoxies. They show satisfactory affinity with polyesters due to the presence of ester groups and, similarly to polyesters, vinyl esters required a catalyst and accelerator for curing at room temperature. Vinyl esters satisfactory withstand chemical corrosion due to less reactive sites and exhibit an acceptable behavior with respect to absorption and hydrolytic attack. Due to these reasons, they represent an attractive choice to be used in boat hulls.

A vinyl ester-based skin on a glass/polyester laminate is able to prevent the osmotic blistering induced by hydrolysis. Boatbuilders or owners can adopt this protection instead of the more expensive epoxies, even if these are a more effective barrier against humidity. Vinyl esters are more flexible and have higher hardness than polyesters. Therefore, if used as a matrix in an FRP laminate, they better withstand the effects of fatigue causing failure in hulls and decks. The vinyl esters have a relative higher working temperature, up to approximately 200 °C. British Columbian West Bay SonShip first promoted this transition replacing polyester resins with vinyl esters for the production of GRP components in their motor yachts. Vinyl ester is increasingly used for the construction of larger ships, including military ships, due to its resilience. Swedish Visby class corvettes have a vinyl ester/carbon sandwich hull with PVC core. On the other hand, the processing of vinyl esters is less easy more challenging if compared to polyesters since a careful preparation of the surface and specific environmental conditions during treatment are mandatory to achieve satisfactory adhesion between fibers and resin. Furthermore, today, resins must have a low styrene content to meet the more restrictive laws on industrial emissions.

Historically, FRP boats have been built using hand layup or spray-up processes. Alternatively, builders have produced high-quality composite hulls adopting pre-impregnated fabrics. This method has higher costs but shorter production times and reduced wastes. The mentioned methods generally consider polyester (PE) and vinyl ester (VE) resins to realize boat structures. These materials offer good water resistance, short cure times and are easy to laminate. However, PE and VE exhibit high shrinkage (7–10%), and high levels of styrene emission that require dedicated systems to minimize worker exposure. Today, the development of novel close-mold and infusion processes is pushing the use of advanced epoxy resins to produce hulls. Epoxies, indeed, provide significant improvements in the quality and performance of the boat and eliminate dangerous emissions, as prescribed by government regulations of the United States and the European Union.

Epoxy systems offer also technical advantages to shipbuilders. They have lower viscosities and can be cured at low temperatures being particularly suitable for infusion processes without autoclave curing. Epoxies have higher elongation, tensile strength, and modulus than PE and VE. Therefore, designers can increase fiber content and decrease the number of layers of the laminates without compromising the mechanical behavior of the hulls. The achieved weight reduction guarantees higher speed and reduces fuel consumption. Furthermore, due to the reduced cure shrinkage (less than 2%), smooth surfaces can be achieved directly on the mold. Many producers, such as Huntsman and Scout Boats, Inc. (Summerville, SC) have benefitted from the use of epoxies. The 42-foot sport fishing boat, manufactured in 2015 by infusion processes, can be considered as an example of weight reduction, improved hydrodynamic performance.

Polyester resins and epoxy resins are widely used as matrices in FPR structures. Nevertheless, significant progress has been reported with phenolics. Phenolic resins are particularly notable for their fire resistance, as they preserve resistance at relatively high temperatures (approximatively 200 °C), with low emission of smoke or toxic fumes. They can be processed by a variety of methods, including hand lay-up, hot-press, infusion. As a counterpart, their mechanical properties are 10–20% lower than those exhibited by polyesters.

Reinforcements

Glass fibers are the most commonly used in large composite structures, excluding some applications in the aerospace sector. Depending on the type of glass, filament diameter, fiber size and chemistry, a wide range of properties and performances can be achieved. Table reports the physical and mechanical properties of the different types of glass fibers available on the market. The most common reinforcing material in the marine application is still E-glass fiber, which has a good maximum tensile strength, about 2200 MPa, and an ultimate tensile strain of about 2.5% combined with outstanding resistance to moisture and chemical aggression. E-glass, due to its chemical composition has excellent electrical insulation properties. Glass fibers with high-strength glass are generally known as S-glass in the United States, R-glass in Europe and T-glass in Japan. S glass has a higher content of SiO_2, Al_2O_3, and MgO and is generally 40–70% than E-glass. Both the

E-glass and the S-glass lose up to half their tensile strength when the temperature increases from the 70 °F (room temperature) to 1000 °F (approx. 540 °C), even if both fiber types still preserve a good resistance.

Table: Properties of reinforcing fibers.

Material	Density (g/cm³)	Fiber Diameter (μm)	Young's Modulus (GPa)	Tensile Strength (GPa)
Carbon HM	1.8	7–10	400	20–2.8
Carbon A	1.9	7–10	220	3.2
Boron	2.6	130	400	3.4
E-Glass	2.5	10	70	1.5–2.0
S-Glass	2.6	10	90	4.6
KEVLAR 49	1.45	12	130	3.6
polyethylene	0.97	12	117	2.6

Carbon fibers have a higher strength and stiffness than glass. Carbon fibers have a maximum tensile strength of about 4000 MPa and an elongation at break of 0.9–2%, depending on the type of carbon fiber. However, the cost of these fibers greatly exceeds that of glass fibers, therefore structures made entirely of carbon fiber are not easily affordable and ship designers are finding very captivating the use of hybrid laminates, made of carbon and glass fibers. Nevertheless, carbon fiber composite structures are more convenient from a design point of view and can be also cost-efficient considering the overall cost (raw material and manufacturing cost). Indeed, the amount of reinforcement can be reduced so that the higher price of carbon fiber can be balanced, without decreasing the performance of the structure.

Fiber-Reinforced Composites in the Offshore Sector

The primary function of an offshore installation is associated with the extraction of crude oil from the subsoil, with the minimum possible impact on its morphology and the utmost respect for safety and environmental protection requirements. The extracted oil is then transported up to the plants along the coast and treated before being sold. Due to the high costs of offshore buildings, it is of fundamental importance to reduce and simplify installations in the open sea as much as possible. Furthermore, the design and installation of new production plants in deeper waters, starting from 300–400 m up 2000 m, cannot rely on the standard technical solutions adopted for shallow waters. Researchers and engineers in the oil and gas industry are abandoning rigid fixed structures in favor of floating structures (still anchored to the extraction structure built on the seabed). In this case, the large displacements allowed in providing the necessary adaptation to wave motion. Moreover, in order to minimize the plant requirements and the costs in the construction phase, builders are promoting the prefabrication of several elements limiting offshore construction work as much as possible. The oil and gas industry started looking at advanced composite materials as a feasible and cost-effective alternative to metals, due to enhanced corrosion resistance, high stiffness and strength and lower weight, when compared to. Typical applications of FRP in this sector include rigid pipes, wound pipes, rigid and flexible risers, structural repairs and more.

Application of Composites in the Oil and Gas Industry

Fiberglass is the most widespread material used to produce composite pipe to transport oil or natural gas from the off-coast drilling station to the onshore plants. The usage of GRP pipes is not so common, however, when the transported fluid is high-pressure natural gas. The fiberglass tube is generally manufactured using two production methods: filament winding (manual or automated) and centrifugal casting. In the filament winding process, the inner diameter of the tube is constant depending on the size of the mandrel, while the outer diameter varies according to the amount of resin and reinforcing material wrapped around the mandrel itself. The orientation of the fiber filaments for the various layers of the pipe needs to be carefully designed to improve lightness without reducing the resistance.

In the centrifugal casting process, resin and reinforcement material are placed inside a rotating mold. The centrifugal action compacts the resin/fiber layers on each other. Afterward, the resin polymerization is activated providing and the tube is formed. Using the centrifugal casting method, the inner diameter of the tube can vary with the amount of resin and reinforcement material used, while the external diameter is fixed. This method produces a high-quality tube, but it requires intensive work and is not competitive in terms of cost. Fiberspar Line Pipe and composite reinforced linear tube (CRLP) represent two relatively new types of composite tubes used in the transport of natural gas. CRLP consists of steel tubes coated or wrapped in a shell of continuous composite material which adds strength and protection to the steel. Fiberspar Line Pipe is a glass fiber reinforced epoxy laminated tube. This tube is currently made in diameters from 1-1/4 to 4 inches and pressure values ranging from 750 to 3000 psi (approximately from 50 to 200 bars).

Steel pipes used to transport oil and gases are highly sensitive to corrosion and subsequent failure. Hydroxides and chloride ions in submerged conditions and in seawater environment accelerate processes leading to metal leaks, cracks, and breakage. Damage evolution is then accentuated by the considerable pressure acting on this kind of component. Being corrosion and loss of metals the main cause for damage in offshore pipes, their repairing is of paramount relevance to the oil and gas industry. As a consequence, intensive study has been devoted to this topic during the past decades all over the world. Conventional repairing techniques for a damaged pipe consist of removing the damaged portion (or the entire pipe) and replacing it with a new one. As an alternative, the damaged region is covered using a layer of welded steel. In other words, external steel sleeves are welded or fixed to the outer surface of the tubes. Welding steel is a cumbersome process especially when it is performed underground or underwater. So, a significant effort has been directed toward the identification of effective and safe repair solutions. The costs and technical issues for repairing and maintenance strategies increase considerably with the operating pressure and the position of the tube damage.

Therefore, relatively lighter and more easily applicable alternative materials are desirable. Reinforced fiber composites have been proposed as an optimal alternative solution for the repairing of these tubular structures, due to their lightness, high strength and stiffness, good corrosion resistance and excellent properties when subjected to fatigue loading conditions. Currently, composite materials commonly adopted for such interventions include E-glass, aramid or carbon fiber as reinforcement and thermosetting resin (polyesters, polyurethanes, phenolics, vinyl esters or epoxies) as matrix material. Depending on the specific pipe repairing method, the aim is to prevent the further extension of the corroded zone, restore the load-bearing capacity of the damaged pipe, avoid fluid leakage from the damaged area, or a combination of them.

Two methods, based on a wet layup or on the usage of pre-cured layers, are generally adopted to repair faulty pipes. Wet layup consists of repairing the damaged area using fibers that are impregnated directly in situ. Special epoxy resins, capable of underwater polymerization, are used as matrices. An alternative method involves the application of epoxy-based resin systems and carbon fibers altogether with a flexible shell. Carbon fibers are circumferentially oriented providing enhanced strengthening to the support steel tube if compared to E-glass reinforcement, assuming the thickness of composite shell as fixed. The pre-cured layer technique consists of using pre-impregnated composites avoiding issues attributable to an underwater resin cure in a restricted and not easily accessible space. This method is mainly applied in case of damage localized on straight pipe sections.

The aforementioned methods can be combined or hybridized in order to (at least partially) solve the drawbacks affecting them singularly. An example is the use of half-shells of carbon to be impregnated in situ rather than using the classic scheme of overlapping sheets of fiber layer by layer. To date, due to the practical challenges in the application of FRP repairing material, metal sleeves still represent the most adopted solution.

The application of FRPs in the offshore sector has been also considered for the production of a variety of primary and secondary structures. Primary or critical structures installed on floating plants include risers and tendons. Secondary structures on floating platforms that can be manufactured from composites include helicopter bridges, piping systems, dwellings, walkways, and stairs. Risers and tendons are flexible systems connecting the floating platform to the subsea structure. Such platforms, in case of significant seabed depth, are generally built converting disused oil tankers, due to the reduced realization cost and the possibility to store crude oil. In figure a scheme of a floating structure Tension Leg Platform (TLP) with risers and tendons is depicted.

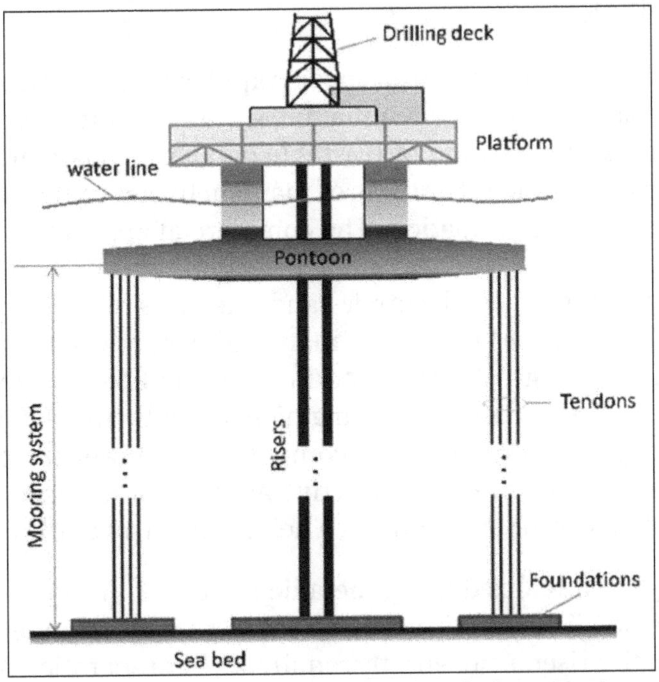

Schematic diagram of the tension leg platform (TLP).

The riser is a tube connecting an offshore floating facility or a drilling rig to an underwater drilling and production systems. Risers are considered the most critical components in offshore pipelines, being exposed to high static and dynamic loads in severe and hostile working conditions. The length and the flexibility of the connecting tubes are crucial to withstand the relative displacements between the submerged structure and the floating station, induced by the actions of wave motion and sea currents. Furthermore, risers act as the framework for the piping elements conveying the crude oil from the wells to the surface plants and as a container in case of oil leakage. Loads acting on the riser are very different in nature (concentrated rather than distributed, static rather than dynamic) and not homogeneously applied along its length. Concentrated forces and torques act mainly at the junction between the riser to the floating platform. On the other hand, distributed loads are attributable to the weight of the riser and (internal and external) fluid-structure interaction. Apart from the gravity and buoyancy action, riser materials are typically subjected to axial traction force on the upper part of the upright, radial solicitation caused by the water current, shear stresses induced by the axial (vertical) flow of the inner moving fluid, concentrated forces reacting floating platform movements.

Several attempts have been made since the 1980s to replace steel risers with FRP ones. Pioneer study to test composite offshore tubulars was carried out by Institut Francais du Petrole (IFP) and Aerospatiale in the 1980s. The performed analysis demonstrated that the closure and suffocation lines represented 30% of the total mass of the riser. Their replacement using composite lines lowered the total mass by 24%, with the subsequent reduction of the dynamic forces acting on the riser. Furthermore, the weight reduction mitigates also the effect of the sea current load. In the mid-1990s, the US Department of Commerce launched a relevant program to develop, produce and test large composite structures, through the investments of the Advanced Technology Programs (ATP) supported by the National Institute of Standards and Technology. Two projects were funded to investigate the application of FRP risers in deep seawater, focused on drilling and production risers respectively.

Despite the differences in the requirements of specific application, the projects presented common elements, aiming at the assessment of the reliability of analysis methods for predicting failure modes and limit conditions, the creation of trustable databases reporting static and fatigue performance, the demonstration of the robustness of manufacturing routes, the definition of metrics for production planning and cost evaluation. The commercial application of the composite risers can be dated back to the 2000s. Nowadays, safety considerations exclude the application of steel risers to depths exceeding 1000 m. On the other hand, the mass of FRP risers is considerably lower if compared to steel (or titanium) ones. As a matter of fact, advanced composites are the only suitable riser material at depths greater than 1500 m. Carbon fibers represent the most common reinforcement material in these structures, having high strength and deformation modulus. They are often combined with glass fibers in hybrid composites, to lower the costs. Aramid fibers are also used. Reinforcement architecture is not strictly longitudinal. Fibers are aligned in both longitudinal and circumferential directions, following a helix pattern, over the entire length of the riser.

Risers have traditionally been realized using metallic materials. As a consequence, severe corrosion damages due to the hostile working environment were very common. For this reason, when the replacement of the entire riser is not strictly required, riser reparation using composite materials is recommended; similar to what happens when dealing with concrete and steel structures restoration. Several works have shown that polymer-based fiber-reinforced composites can be used effectively for repairing steel pipe. In his study, C. Alexander investigated the use of E-glass and carbon fibers composites with urethane and epoxy resins in different repair strategies. The highest tensile load-bearing capacity and axial stiffness were achieved using unidirectional carbon fibers with epoxy resin. As a counterpart, this solution was the most expensive one.

The choice of materials and manufacturing strategies are very case-specific, depending on the trade-off between mechanical properties and economic convenience. In principle, glass, carbon and aramid fibers can be used for repairs. Flexible wet layup and the pre-cured layered systems are successfully used in riser repairing. However, these composite repair strategies present similar limitations as discussed in the repairing of submerged pipes.

Apart from risers, floating platforms (preferred in the case of depths major than 300 m) are supported and held in place using the so-called attack group. It is constituted by tie rods, ropes, and tendons. These components are realized using composite materials, providing an easily transportable and installable solution, using regular ships instead of heavy lifting crane ships. The deepest TLPs measured from the seafloor to the surface are:

- Magnolia TLP (1425 m).
- Marco Polo TLP (1300 m).
- Neptune TLP (1300 m).
- Kizomba A TLP (1177 m).
- Ursa TLP (1200 m).
- Allegheny TLP (1020 m).
- W. Breast A TLP (1000 m).

The body of the connecting unit is formed by parallel wires manufactured pultruding carbon fiber bars that guarantee great flexibility, high deformation modulus, and resistance. Each group is made of 50–100 truss. The final size of the group is determined by the number of threads, depending on the strength and stiffness required for the particular TLP. Even though the cost of carbon fibers is significantly decreasing since their introduction in the 1970s, their use is still not economically justified in structures that require many and long tie rods. It is worth noting that for a single TLP, installed on 5000 feet deep water, an anchoring system made of carbon fiber composite requires approximately 2.2 million euros of fiber. The need for cost reduction preserving the load-bearing capacity has promoted the adoption of hybrid composites based on carbon and aramid fibers as reinforcing materials. Epoxies are the most common resins adopted in TLP tendons.

Caissons are large-diameter pipe transport systems suspended from offshore platforms and used for the intake and distribution of the seawater, necessary in various processes and utilities, such as heat exchanger, feeding the desalination and purification systems, transportation of waste products and fire applications. Their size ranges from 250 to 1200 mm in diameter and from 20 to 100 m in length, depending on the function and the size of the platform. The standard caisson includes a pumping and rising station to bring seawater to the platform as well as a series of smaller tubes that discharge the treated fluids. Steel caissons are oversensitive to corrosion.

Consequently, they require rigorous and continuous inspections, maintenance and repair or replacement procedures to ensure compliance with safety laws. These procedures, often including underwater inspections, are extremely expensive and their cost can reach up to five times the cost of the original installation. Numerous offshore suppliers have considered FRP, as a suitable alternative to steel for caissons production, whereas. The use of GRP materials for caisson offers corrosion resistance, reducing the cost of maintenance, and a 50% weight saving compared to traditional steel caissons, making also easier their handling and installation.

Materials

Steel is the most commonly used materials in the oil and gas industry thanks to its considerable specific weight and good resistance to towing but it is excessively sensitive to corrosion. Therefore, in order to ensure the durability of offshore components, avoid losses of mechanical resistance connected to the chemical decay, steel pipes have to be equipped with an efficient cathodic protection system by means of a sacrificial anode. Subsea pipelines, due to external pressures and bending stresses during laying, must be made with steels with high mechanical strength (type API 5L class X65 and above).

When pressure is the critical factor steel is preferred to fiberglass composites and plastics. The pressure values of the steel pipe can exceed 5000 psi (i.e., 350 bars) depending on the size of the pipe and the location of the installation. High-density polyethylene (HDPE) can also be used to manufacture tubes for transportation of natural gas and oils due to the outstanding chemical resistance. The use of high-density polyethylene (HDPE) pipes to transport natural gas is limited to pressures minor than 100 psi (7 bar). Fiberglass/epoxy FRP has been tested under pressures higher than 3500 psi for the tube with diameters ranging from 2" to 3" and about 2750 psi for 4-1" diameter tubes. The vulcanized epoxy tube was successfully certified at 3750 psi for 2" and up to 1100 psi for the 8" diameter tubes. Thermoplastic pipes are susceptible to a reduction in mechanical properties at high temperatures.

In the case of HDPE pipes, the in-service temperature should not exceed the limit of 100 °F (i.e., 37 °C). Mechanical properties of FRP material also decrease with an increase in temperature, but not as dramatically as HDPE. Pressure glass fiber tubes are certified up to approximately 200 °F. Fiber-reinforced vinyl-ester tubes retain their properties up to around 300 °F but in general, only steel tubes can withstand high temperatures. Cast iron, plastic materials (e.g., high-density polyethylene, PEAD), glass fiber reinforced polyesters and cementitious materials are also frequently used. Concrete pipes are heavy and guarantee good resistance to abrasion, but they are difficult to be joined with each other and are sensitive to corrosion of the reinforcement (steel bar). The cast-iron pipes have the same pros and contra as steel pipes, with further issues related to the joining.

Reinforced polymeric composites are considered an attractive alternative over conventional materials. This is particularly related to the superior corrosion resistance, high strength to weight ratio, and flexibility. The ability to tailor the mechanical properties controlling the type, quantity, and direction of the reinforcing material makes the composite tube an ideal candidate for applications in high-pressure conditions. On the other hand, they must be weighed down in subsea applications (e.g., submerged pipes) and have a low resistance to abrasion.

The used resins are both thermosetting and thermoplastics. Thermosets have found widespread use in offshore oil and gas, even if they typically exhibit more limitations than thermoplastics.

Indeed, even if thermosets are relatively cheaper in terms of raw material and manufacturing, their shelf life is significantly limited. Furthermore, they offer higher mechanical properties and thermal stability than thermoplastics. On the other hand, the latter show better chemical resistance and lower water absorption, which makes them particularly suitable for offshore usages, combined with a notable recyclability. A qualitative comparison between typical thermosetting and thermoplastic resins is reported in table.

Table: Properties of thermoplastic used in oil and gas applications.

Material	Properties	Notes
PA11/PA12	Moderate thermal resistance Resistant to CO_2	Maximum temperature 80 °C. Not resistant to hot water, acids, bases or H_2S at temperatures higher than 65 °C.
PE/HDPE/ PEX	Good chemical resistance Resistant to CO_2 and H_2S	Maximum temperature 90 °C. Not resistant to aromatic substances or hydrocarbons at temperatures higher than 65 °C.
PVFD	Good chemical and thermal resistance Resistant to CO_2	Maximum temperature 130 °C. Not resistant to bases.
PEEK	Resistant to temperatures over 330 °C Good fire-retardant capacity and chemical resistance	Expensive High viscosity.
PP	Affordable cost Good chemical properties Good workability	Maximum temperature 110 °C. Not resistant to aromatic substances.
POM	Good mechanical properties Excellent fatigue capacity Exposure temperature limit around 150 °C	Not flame or UV resistant. Not resistant to hydrocarbons above 80 °C.
PPS	Limited exposure time at temperatures around 260 °C Excellent chemical resistance properties Fireproof	Difficult to extrude unless using special additives, limited to 180 °C.

Table: Properties of thermoset resins used in oil and gas applications.

Material	Properties	Notes
Epoxy	Good chemical resistance. Moderate thermal resistance.	Maximum temperature < 100 °C.
Phenolic	Good chemical resistance. Moderate thermal resistance. Good fire resistance.	

Fiber-Reinforced Composites for Marine Renewable Energy

Application of Composites in Marine Energy Production

The category “marine renewable energy systems” includes devices and plants installed in the marine environment and harvesting the energy carried by waves, currents, tides, salinity, and temperature gradient as well as wind. Among these, tidal stream energy, wave motion and offshore wind represent the most mature technologies, where glass and carbon fiber reinforced composites find increasing use. The thermal energy conversion (OTEC) is based on the temperature difference between the surface and the deep water to produce electric energy from a thermal cycle. Tropical areas are more favorable for this method as higher temperature gradients are reached.

The osmotic potential derives from the difference in salinity between seawater and freshwater in continental areas. Its use is possible where the freshwater flow into the sea is very abundant (as in the Baltic Countries in Europe). These last two concepts are still at a theoretical planning stage.

The wave energy derives from the difference in gravitational potential energy between the cable and the crest. To date, the existing wave energy technologies are divided into various categories based on the position with respect to the coast (shoreline, coastal area, wide) and on the working principles. Based on the method used to capture the energy of the waves, devices are classified as follows:

- Oscillating water column (OWC): It is a partially submerged structure with an integrated air chamber, which funnels the waves inside below the waterline. Due to the oscillation of the fluid, the trapped water column rises and falls alternately pressurizing and depressurizing the air column and pushing or pulling it through a turbine. It can be located onshore or offshore in deep waters.

- Overtopping devices: They consist of partially submerged structure collecting the seawater in a reservoir at a level slightly higher than the surrounding sea. The system uses the potential energy stored in the trapped water that runs back to the ocean through a low-head hydraulic turbine. Devices can be either onshore or floating offshore. Floating devices present some environmental concerns coming from the mooring system that could affect benthic organisms or from electromagnetic field (EMF) effects produced from subsea cables.

- Oscillating wave surge converter (OWSC): It consists of floating plates having one end to the fixed to the seabed and connected to hydraulic pumps and a turbine. The system uses the oscillation of the plates with respect to the fixed point to directly capture the wave energy without using a reservoir or a collector. The waves cede its mechanical energy to the floating plates that in their movement compress the hydraulic pumps, in turn, the pumps transfer the energy stored in the internal fluid to the turbine rotor.

- Point absorber buoy system: It uses the lifting and lowering of a floating object during the passage of the wave to drive hydraulic pumps and generate electricity. The devices float on the surface of the water and are held in place by a cable connected to the seabed. There are also submerged models, namely submerged pressure differential. In this case, the motion of the waves causes the rising and falling of the sea level inducing a pressure differential in the submerged buoys. Their continuous oscillatory movement drives a fluid pump to transfer the mechanical energy.

- Attenuator: This device acts in a similar way to the buoy system, with multiple floating segments connected to each other and oriented perpendicular to the waves. The relative oscillation of the floating segments converts the energy from incoming waves driving hydraulic pumps to generate electricity.

Energy from tidal currents is generated by daytime or semi-diurnal regular cycles. One of the main advantages of tidal energy compared to other renewable energies, such as solar and wind energy is that energy production is predictable on long-term scales due to the well-documented behavior of the tides. Furthermore, the tidal waves are much more constant than the wind; therefore results are more quantifiable a priori. Due to the high density of seawater (at least 800 times than that of the air) submerged turbines undergo significant workloads with respect to the wind turbine. Furthermore, the used materials must preserve high levels of strength and stiffness and withstand the corrosive seawater environment, and require few or no maintenance, considering the inaccessibility of deep seawater area. For these reasons, fiber-reinforced polymeric composite materials, especially GFRP and CFRP, represent an attractive choice for the realization of tidal turbine blades. GFRPs are generally preferred by designers due to the trade-off between structural properties and low-cost. CFRP offers high performance and lower weight (by about 13% for the blade) if compared to GFRP but at a 7/8 times higher cost. Due to this reason, the use of carbon fiber composite is limited to the main parts of the blades.

Differently from wind turbines, horizontal axis tidal turbines are affected by issues such as cavitation in blade tips and higher blade bending moments. In particular, GFRP blades during its lifetime (approximately 20 years) is subjected to an average speed of tides that varies between 2.4 and 4 m/s, resulting in early damage due to the strong bending moments. These extreme working conditions impose an upper limit on the dimension of the blades around 12 m. Conversely, CFRP provides a significant improvement in the maximum bearable stress, being the loads well below the traction (axial) and shear failure stresses. It is reasonable to conclude that GFRP is not a suitable material for the main structural components of large submerged turbine blades.

Submerged turbines are difficult to access for monitoring and maintenance, therefore the blades are designed in order to withstand the aggressive seawater environment for a long time. Besides the common issue related to fatigue loading, galvanic corrosion phenomena, as well as fouling and damages caused by the presence of marine organisms, need careful consideration. The effect of the exposure to the marine environment on different materials, commonly used in high-demanding applications (automotive, aeronautic, etc.), namely:

- Glass fiber in an epoxy matrix.
- Carbon fiber in an epoxy matrix.
- Aeronautical aluminum alloy.
- 314 and 316 stainless steel.
- Common steel.

After one year of exposure, composite blades showed a very low level of fouling on the surface, while the metal components resulted affected by severe corrosion, except for ones manufactured in stainless steel. Wind energy production does not generate any solid, liquid residues or gaseous emissions, nor involves the depletion of any form of fuel. A few concerns are related to the noise and visual disturbances, or eventual impact on flora and wildlife. These factors represent an obstacle to installation in proximity of inhabited centers, but they are not critical in several locations. Obviously, harvested power is hardly predictable and not constant, as it is dependent on wind speed. Offshore wind energy plants are exposed to higher and generally constant wind speeds, with the possibility of extending the plant over large areas without incurring in restrictions related to the environmental impact.

However, the design of foundations and support structures, similar to the ones used in offshore oil and gas plants, and of turbines and blades exposed to the corrosive marine environment, has to meet additional constraints with respect to the in-shore wind farm. Therefore, offshore wind energy parks present relatively higher costs in terms of materials, installation, maintenance, and connection to the grid. Recent developments in new technologies, advanced material, and manufacturing processes are making the realization of offshore wind plants more affordable and convenient. Some studies pointed out that the cost of energy required for installation, the useful life and the disposal of an offshore wind plant is typically recovered in 6–12 operative months. Considering a standard lifetime of approximately 20 years, the ratio between produced and consumed energy (efficiency factor) oscillates between 20 and 40. The study conducted by Vestas Wind Systems revealed that the energy used in the production, installation, and disposal of a modern 3 MW (MegaWatt) plant should approach the break-even point in about 6.6 months for ground installation or 6.8 months for an offshore plant. Assuming a useful life of 20 years, energy efficiency is about 35 for both cases.

Blades are usually manufactured as multi-material structures, combining reinforced fiber polymers and wood in order to meet strength, stiffness and lightweight requirements. Reinforcements are typically glass, carbon, or hybrid fabrics. Polyester, vinyl ester or epoxy resin is used as matrix. As for tidal turbine blades, upper and lower parts are produced separately and then joined each other by mean of the adhesive layer, resulting in a sandwich structure. Adhesive strength and durability are crucial parameters since they significantly affect the performance of the blade.

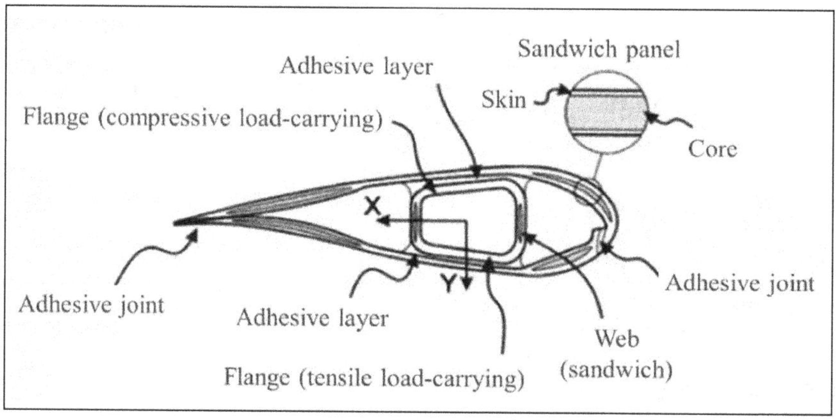

Cross-section of a wind turbine blade.

Materials

Carbon and glass-reinforced composites represent a suitable replacement for metals in the manufacturing of blades for offshore wind and tidal energy devices. As a matter of fact, FRP provides high specific strength and stiffness coupled with outstanding resistance to corrosion, moisture, and fatigue. Furthermore, they can be molded into complex shapes with reduced overall mass. Design and manufacturing of composite blades for wind turbines and for on-shore and offshore devices have the same requirements. Nevertheless, tidal turbine blades, being immersed in the seawater environment, require more attention at the design stage, for the definition of the shapes and the selection of materials. Indeed, tidal turbines have to withstand both extreme static and fatigue loads during their lifetime (approximately 20–25 years, like the wind turbines). The fatigue loads depend on the depth of the installation, the proximity of the seabed and the wave motion.

Consequently, they are extremely variable and hardly predictable. Intensive experimental and numerical studies have been carried out to investigate the effects of cyclic loading (in particular mean stress and variable amplitude of applied loads) on the degradation of composites in the seawater environment. Furthermore, tidal turbines must be designed in order to not avoid maintenance during their lifetime, because of the inaccessibility of the operating environments (usual turbines are placed at a depth of about 50 m where sea currents may reach extreme speeds of about 4 m/s). Finally, high resistance to biodegradation is required to withstand the hostile seawater environment.

The adoption of carbon and glass as reinforcement in composites is driven by a trade-off between material and manufacturing costs, in order to make the produced energy competitive and meet design requirements (e.g., blade hydrodynamic profiles). The need for cost-minimizing pushed the designers towards the use of carbon fibers only for the realization of the primary load-bearing parts of the structure. Epoxy and vinyl ester resins are commonly used as matrices. The former exhibits superior performance compared to other resins, especially in cyclic loading conditions. The latter is less expensive and is characterized by higher resistance to water penetration. Carbon fiber reinforcements are being introduced into the blades to improve stiffness and tensile strength, compared to glass-fiber composites.

However, the improvement in compression strength is considerably lower compared to the rise in overall costs. Therefore, using a mixture of glass and carbon fibers, with carbon used primarily to increase stiffness, represents a compromise between these competitive goals. Tests carried out on epoxy and vinyl-ester resin laminates with E-glass reinforcements showed that epoxy/E-Glass panels have 25% higher fatigue resistance than vinyl ester/E-Glass panels, in the case of in-plane fatigue loads on dry samples, without considering porosity, holes, and ply-drops. Other studies showed that the static properties of glass-reinforced vinyl resins exhibit less degradation after aging in seawater than the one that epoxies. A further concern about the use of FRPs is their poor recyclability. Thermosetting resins cannot be recycled and the composites containing these resins can be disposed of only by incinerating. Thermoplastic resins, on the other hand, offer a high hardness combined with a greater degree of recyclability. However, blade production requires high working temperatures and the bonding of finished pieces could be difficult. Today, environmental concerns are pushing the development of highly recyclable materials such as natural cellulose fibers and bio-based polymers.

Carbon fiber-reinforced composite proved to be the material that better withstood the marine environment, carbon fiber composite showed after 10 months the lowest percentage of dirt sedimented on the external surface of the components. In addition, tests showed that surface finishing of the materials affects the percentage of organisms stuck on the surfaces of the rotor, it is higher in case of rough surfaces or cracks promote the formation of fouling, conversely, it remains low in case of smooth surfaces. Results also indicated that the biological contamination is cumulative, when a surface is affected by a certain degree of dirt or corrosion, it becomes more susceptible.

GFRPs are more susceptible to corrosion and biological contamination than the CFRPs, however, they are being still used due to cost reduction strategies. To solve this problem, post-manufacturing processes, such as the electrolytic deposition of Ni-P, are carried out. These increase the surface finish, reducing the porosity, the surface hardness and the corrosion resistance of the material, protecting it from chloride ions.

First large-scale turbine blades in fiber-reinforced plastics were manufactured by hand layup of pre-preg. The main parts of the tidal turbine blades (i.e., upper and lower skins and box-section spars) were produced separately and then joined using adhesive at very high production costs. Resin infusion processes have been employed to minimize costs allowing for the reduction of the number of parts to be assembled and the increase of process automation. Resin infusion has been already applied in the production of wind turbine blades, but the adhesive bonding of the upper and the lower skins of the blade is still necessary to achieve the final product. The "Integral Blade" process, patented by Siemens, represents an exception, it uses a closed external mold and a flexible and expandable inner bladder to produce at the same time the entire glass/ epoxy blade.

Airborne Composites Company is employing a similar technique to manufacture tidal turbine blades. They produced four meters long blades, installed in the 500 kW CoRMAT turbine, developed by the European Marine Energy Centre (EMEC) since 2014, using a VARTM 'One-shot' process. The VARTM "one-shot" process ensures higher production efficiency, eliminating the presence of additional material around the interface of the connectors. The company has also developed a method to incorporate inserts at the root of the blade and to connect it to the hub before the resin infusion. As a consequence, drilling operations on the realized composite are removed and the thickness of the laminate is reduced. Some issues related to the quality of the laminate and the presence of defects in the manufactured parts is still unsolved. For single laminates and sandwich structures, manufacturing defects include delamination, dry areas, and un-wetted fibers, porosity, wrinkles, defects in fiber reinforcement, fiber misalignment.

Non-destructive testing represents the most effective technique to detect manufacturing defects for wind turbine parts, due to the large areas to be monitored, complex shapes, the difficulty of accessing inner parts and the variety of detectable defects. Relevant advances have been made in non-destructive testing of FRP in recent years, however, major challenges are still present. Currently, visual and ultrasonic techniques are the most used methods, even do not provide a comprehensive and exhaustive inspection. Therefore, combinations of different control techniques are typically adopted. Besides the defects resulting from the manufacturing processes, it is of paramount importance monitoring the decay of turbine blades during the working life (i.e., fatigue damage evolution) and detecting the incidence of damages caused by accidental events such as lightning and hail storms. Structural monitoring systems are being developed, based on the analysis of acoustic emission or use of optical fiber sensors, allowing the continuous tracking of the state of degradation of materials and structures and the prediction of the residual useful life by means of software and models for damage assessment. Detection of unexpected changes in the state of degradation may be used to start the inspection, repairing or replacement of the rotor blades. The advantages of implementing a failure detection system using advanced sensors are summarized as follows:

- Prevent premature and irreparable breakdowns or secondary breakdowns.
- Reduce maintenance cost: On-line inspection allows extending the time interval between inspections and avoiding extraordinary maintenance with the replacement of still working parts.
- Remote supervision and diagnosis, useful for large turbines installed in remote and not-accessible sites (open sea).

- Improvement of the capacity factor: Thanks to the early warning of unexpected damage, restoration activities can be carried out during optimal wind conditions.
- Support for the design of the next-generation wind turbines.

Underwater Structures Repairing

Repairing of underwater structures, such as concrete or metal pillars, represents one of the most common and interesting applications of FRP in the marine environment. Indeed, confinement and reinforcement are achieved easily and efficiently by wrapping the structural elements with fibers or fabrics impregnated with resin. Initially, FRP was used mainly for the restoration of pre-compressed concrete structures that experienced severe corrosion. Afterward, these materials proved to be well suited for repairing even wooden and steel pillars. As well known, corrosion is the main cause of concrete degradation. The concrete is an alkaline substance with a pH in the range of 12.5 to 13.5. This high alkaline environment protects steel rebar, promoting the passivation of steel on the surface. Unfortunately, during its lifetime, concrete loses its protective characteristic. This occurs essentially by three different mechanisms:

- Carbon dioxide reacts with the alkali in the cement making the water inside the concrete pores more acid and lowering the pH. The process destroys the oxide film on the steel rebar surface and hinders the passive conditions. This process, called carbonatation, starts from the surface and gradually moves deeper and deeper to the core.
- In presence of electric fields, crossing stray currents interfere with the reinforcement and the protective film can be destroyed in areas where the current flow out from the reinforcement.
- The concrete gets in contact with marine environments containing chlorides. The chloride ions penetrate into the concrete and reach the reinforcement. If critical chloride content exceeds the surface of the reinforcement, the protective film can break locally. The concrete attacked by chlorides corrodes exposing the reinforcement bars to the action of oxygen and water. The bars attacked by atmospheric agents corrode and increase in volume causing the concrete cover to be expelled.

The marine environment accelerates the degradation process of reinforced concrete generating localized damage in a specific area called splash zone. This area is subjected to continuous drying and wetting due to the tidal cycles and, in addition, to the action of the waves breaking against it. Therefore, the surface of the pillars goes from completely dry to completely wet conditions. Consequently, chemical and mechanical actions promote spalling or delamination of the concrete cover. Figure shows the different damages that could affect the concrete during its life-time.

FRP repairing systems exert two actions. They provide protection against the corrosion (non-strengthening repair) and act as reinforcement for the concrete structure (strengthening repairing). In the former case, the resin, penetrating the porous layers of the concrete, hinders the entry of water, rich in chlorides and oxygen, both in the cement and in the reinforcement, preserving the steel from oxidation. Furthermore, the reinforcing fibers, placed along the longitudinal and the transverse directions, increase the shear and bending resistance and avoid also the enlargement of the pillar section thanks to the compression action. Therefore, the reinforced fibers enhance both the action of the concrete (which works in compression) and the action of the steel rebars (which work in shear and bending).

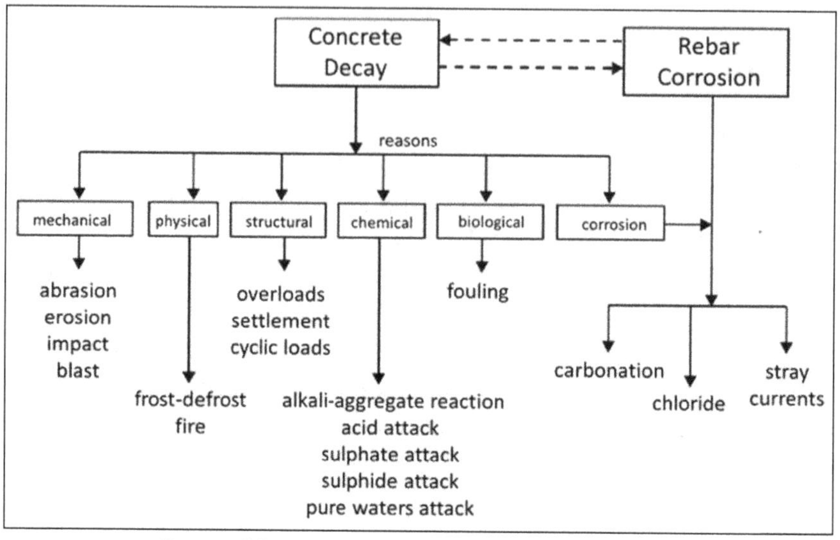

Types of damages occurring on a concrete pillar.

Repairing Techniques

Two different methods are generally adopted, based on prepreg and wet-layup, respectively. The former method requires placing a reinforcement jacket made of pre-impregnated fibers on the surface of the pillar to be repaired. Urethane resins, which can be activated with water, are usually employed. Carbon and glass fibers can be adopted, as reinforcements when the repairing involves the application of a prepreg, otherwise, in case of the wet layup method, only the use of fiberglass is recommended, since the positioning in situ of dry carbon fiber, especially underwater, presents several issues and is excessively time-consuming. Prepreg layers will be placed on the pillar in such a way to have fibers oriented along the longitudinal direction of the pillar and the transversal direction: the longitudinally oriented fibers increase the bending strength, while the fibers in the transverse direction withstand the compressive load and avoid the widening of the pillar section (they enhance the resistance of the concrete pillars against the expansive forces).

Bidirectional fabrics are usually preferred to the unidirectional fibers since they can to offer support against both loads minimizing the numbers of layers required and respecting the space constraints of the repaired structures. Usually, the number of glass fiber composite layers is twice that of the carbon fibers since the latter have better mechanical characteristics. These techniques have been successfully employed in restoration and strengthening of pipelines, bridges and coastal or marine structures, such as the Singapore's Marina Bay Stadium. In that case, a special repairing system was used adopting fiberglass and resins able to polymerize under water. In this way a simple and quick installation of the fabrics was possible, avoiding the realization of expensive temporary structures for the removal of the water. Carbon and aramid fibers can be used as reinforcement.

In the wet layup system, the choice of epoxy resin system depends on the possibility of clearly distinguish the dry from the wet areas on the surface pillar. In this case, two different types of epoxy resins can be used, otherwise, hydrophobic epoxy resin is used on the whole pillar. Nevertheless, due to the cycles of the tides, the second type is often preferred in both conditions. The wet-layup method represents a much more flexible solution since the fiber arrangement can be modified with respect to the standard stacking sequences (e.g., 0°/90° or 45°/135°) used in the prepregs and the size of the fiber fabric can be also easily adapted. However, in-situ impregnation is more time-consuming and more expensive than prepreg method, due to the additional time required by the stacking of the pile of the fabrics, the resin cure and by the transportation of all the equipment necessary for the infusion at the site of the repairing. Tests performed with both methods on the Gandy Bridge in Tampa Bay and on the Allen Creek Bridge highlighted that:

- Repairing using prepregs takes about 30/45 min for each damaged pillar, while the repair through wet layup takes about 90 min.
- The prepreg system gives better results in the case of repairs on mostly dry pillar areas, while the wet layup system gives greater guarantees of success in submerged areas.

The prepreg system, being fast and versatile, can also be adopted in the case of urgent repairs. In order to achieve a satisfactory restoration of the damaged structure, the formation of an adequate bond between the substrate to be repaired and the resin is mandatory. The effectiveness of the repair using FRP is strictly connected to the strength of the bond: a scarce bond causes a poor repair with anticorrosion protection destined to fail in a short time. To this regard, surface preparation of pillars also represents a key factor for acceptable bonding. Piles need to be accurately cleaned from marine organisms deposited on the surface and from debris or fragments detaching from the body of the pillar, in order to ensure an intimate contact between the FRP and the substrate. In that way, clean pores in the cement can be filled with resin. Also, the iron oxide must be removed from the steel rebar surface, to interrupt the oxidation reactions and the progress of corrosion.

Wrapping techniques, namely vacuum bag (or vacuum bagging) and pressure bagging have been developed to guarantee a constant and uniform pressure during the hardening phase. The applied pressure ensures both the bond strength between the reinforcement and the substrate and nearly constant thickness of the reinforcement jacket in each point of the pillar. The vacuum bag is designed to apply pressure on the FRP composite creating a vacuum (the maximum pressure acting on the material to be compacted is limited to 1 atmosphere) without a mold and a counter-mold but using a flexible diaphragm (bag) and air to apply the pressure.

Creating and maintaining the air-tight seal during the curing process is critical because the concrete of the substrate to be repaired typically presents some lesions or cracks from which air takes over inside. Another key factor is the removal of excess resin from the mold. This must be done in low viscosity conditions, so before the resin reaches the gel point, in order to prevent compression loads on the material when hardening has already begun. The main benefits of vacuum bagging can be summarized in the removal of the air trapped between the layers, compaction of the fiber layers providing efficient transmission of forces and preventing the movement of the fibers, moisture reduction, and optimization of the fiber/matrix ratio in the composite structure. The pressure bagging technique consists of a low-resistance air-tight bag housed by an outer rigid or flexible structure. The inner airtight bag is connected to a compressed air source. The outside bag is made of nylon with a vertical metallic closure.

Pressure bagging allows applying a compaction pressure constant and uniform over the entire surface of the wrapped pillar and greater than that in vacuum bagging. In general, both methods are able to enhance the FRP-concrete bond in the dry and submerged regions. Pressure bagging is preferred whenever there is a need to use higher pressures to compact the FRP, such as when using more viscous resin types. When epoxies are used, it is also more effective if combined with the wet lay-up technique, because the relatively higher pressure promotes the resin flow into the concrete pores, improving the bond quality. On the other hand, vacuum bagging is preferable when using polyurethanes, since the gases generated during the polymerization of the resin (such as the carbon dioxide produced by the polyurethane) can be easily removed, avoid porosity formation.

Repair on Wooden Pillars

Concrete is the most employed material for the submerged structures in the marine environment, however, treated wooden poles have traditionally been used for the construction of piers and other structures along the waterfront. The main concern regarding wooden poles is deterioration due to marine organisms. This limits lifespan and imposes frequent repairs and replacements. Since the use of conservation treatments for wooden poles has been reduced due to environmental and human health concerns, there is a current need for efficient methods for protection. Damage of wooden poles in marine constructions derives from physical and chemical actions. The former type is mainly related to the action of marine organisms, fungi and marine worms that inhabit the seabed, which stick to the body of the pole, corroding and deteriorating it. It has been observed that fungi attack the pillar above the waterline, while woodworms are mainly present in submerged areas. So, differently from the concrete pillars (mainly affected the damage in the splash zone), in the case of wooden pillars, the most damaged area is in the tidal zone, immediately below the splash zone, due to the coexistence of fungi and woodworms. Figure below schematizes the level of damage that could occur on the wooden pillar depending on the area of exposition to the marine environment.

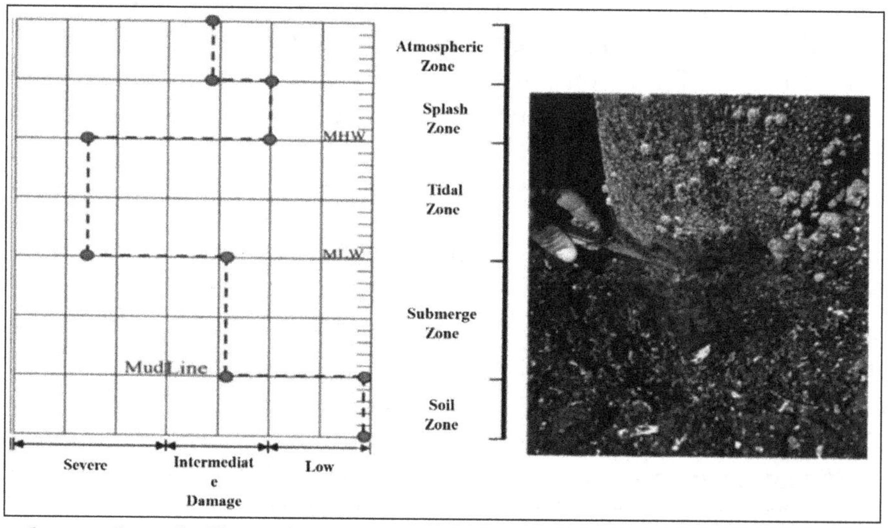

Photo of a submerged wood pillar at the mudline (right) and typical damage profile of wood pillar (left).

The mechanical action of ice, debris, sand, etc., also contributes in reducing the integrity and the lifetime of the pillars. The use of FRP composites is a feasible solution to repair wooden pillars.

Different methods to promote the junction of the substrate with the reinforcing material have been investigated so far:

- Filling the space between reinforcement and pillar with cement which acts as a binder and insulator.
- Using polyurethane foam to isolate the damaged part from marine worms and fill the space between pillar and composite.
- Mechanical joining between pole and reinforcement (threaded steel rods).
- Using cement-polymer coating (epoxy resin + cement).

If compared to the other methods, the cement-polymer coating technique has been claimed to give the best result in most cases, increasing friction between reinforcement and substrate and consequently reducing the relative displacement. Nevertheless, this remark cannot be fully generalized since the variation of loads and environmental conditions can suggest different strategies.

Repairing techniques, especially in the case of emergency repairs, are in continuous evolution aiming to develop new adaptable and versatile methods. To date, several fiber-reinforced composites are now available to achieve an effective restoration of damaged marine structures. As an example, a wearable FRP jacket able to provide support to any pillars is proposed in. This solution consists of a coating that can be installed around a damaged pillar, without using prepregs layers or even in-situ impregnated fibers. High-density polyethylene pipes cut along the longitudinal direction are used and glass or nylon reinforcing fibers arranged in a transverse direction. The polyethylene pipe is very flexible, allowing it to be applied to the pillar without the use of hinges and without incurring fractures of the material. Figure shows the scheme of the reinforcing jacket applied to the wooden pillar.

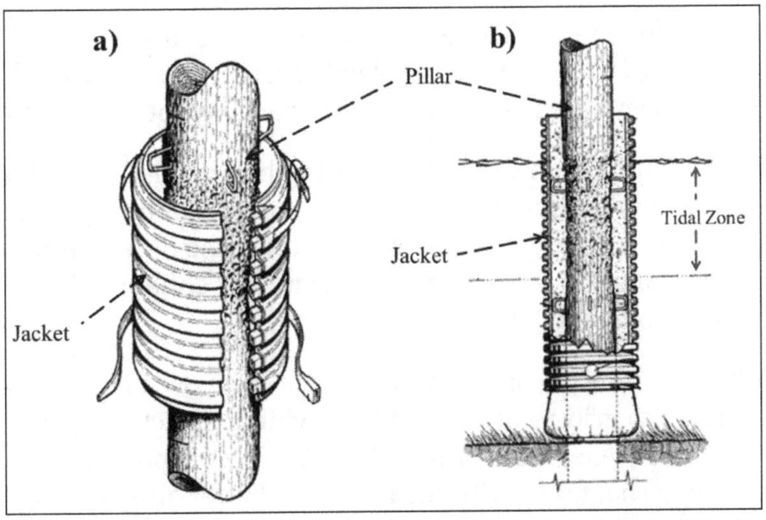

(a) scheme of pillar repair jacket, (b) example of installation of the jacket within the tidal zone.

The coating layer present external undulations, necessary to increase the stiffness and strength of the structure and to facilitate joining and fixing operations. The reinforcement fibers are placed inside the undulations. The junction between the substrate and the reinforcement is guaranteed by concrete, which acts as a filler to repair any cracks on the surface and as a binder. FRP composites are being increasingly used not only as reinforcement or repairing materials but also as a structural element for the construction of pillars in marine environments replacing aluminum, steel, wood, and PVC. Five types of composite pillars are currently used as showed in figure.

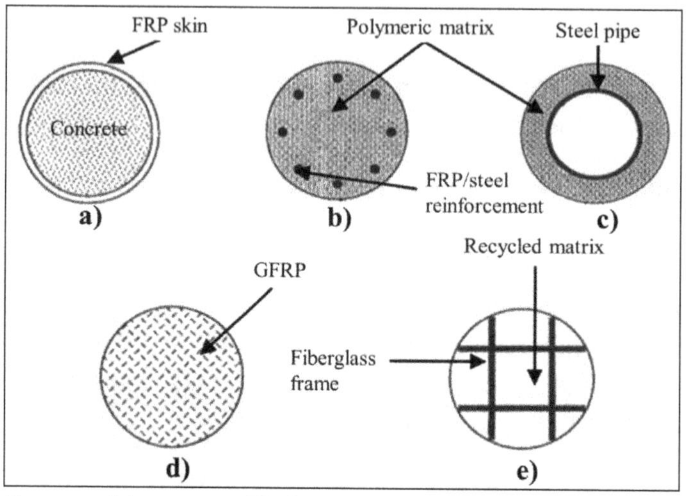

Composite pile types: (a) concrete-filled FRP pipe, (b) SRP pipe, (c) steel pipe core pipe, (d) fiberglass reinforced plastic pile, (e) fiberglass pultruded pipe pile.

- Steel pillars coated with plastic skin, in which the steel core provides the mechanical strength while the plastic cover protects the steel from corrosion. High-density polyethylene (HDPE) from recycled plastic materials is used as a coating. Steel core pipes were used in the port of Los Angeles, in San Diego and in El Sauzal, Mexico, however, delamination issues between plastic skin and metalcore, due to mismatching in the thermal expansion of the two materials, was observed.

- Structurally reinforced plastic pillars realized with extruded plastic matrix and glass fibers or steel bars as reinforcement. The reinforcing elements are arranged in a concentric pattern within the inner plastic core and are sized according to the type of stress to be supported. Plastics outer cover is treated with additives to delay the degradation.

- Concrete-filled FRP pillars, in which FRP outer shell is filled with unreinforced concrete. The FRP offers structural noncorrosive reinforcement, compresses the concrete providing also isolation against an aggressive environment. On the other hand, concrete is responsible for internal resistance in the compression zone and enhances the overall stiffness of the element contrasting buckling of the FRP shell.

- Fiberglass pultruded pipe piles: The outer part is composed of FRP composites sheets fitted out with fiberglass grid reinforcement, providing structural support. The grid inserts are filled with polyethylene, foam or recycled plastic. In fender piling applications, the filling connects the external fibers with the inserts to better absorb eventual impacts.

- Glass fibers reinforced plastic pillars, which involves the use of recycled plastic with randomly arranged glass fibers. Polyethylene and 20% glass fiber are conventionally used. The outer shell is dense and solid, while the plastic core is filled with foam to reduce the total weight. Various additives are also employed to enhance the mechanical behavior of the materials and increase their durability.

Biomedical Applications

The use of natural or synthetic materials to substitute or integrate body functions or organs, damaged by traumatic or pathologic events, to assist tissue healing or to correct abnormalities, dates far back to the beginning of medicine into ancient civilizations. Therapeutic innovations and the design and implementation of complex medical devices permit increased patient survival, significantly improve quality of life and contribute to increase life expectancy.

The word "composite" refers to a heterogeneous combination, on a macroscopic scale, of two or more materials, differing in composition, morphology and usually physical properties, made to produce specific physical, chemical and mechanical characteristics. The combination of different elements results in a material that maximizes specific properties. The advantage of composites is therefore that they show the best qualities of their constituents, and often exhibit some properties that the single constituents do not have. Moreover, composite materials allow a flexible design, since their structure and properties can be optimized and tailored to specific applications.

Cardiovascular Applications

Various congenital or acquired lesions related to the heart and blood vessels lead to severe functional diseases requiring surgical treatments. Cardiovascular devices are therapeutic supports for heart and blood vessels. Thrombus formation is the most critical issue for materials placed in contact with blood and is strongly dependent on surface characteristics, implantation site, duration, local hemodynamics, etc. The development of a perfect hemocompatible material is still an unsolved problem.

At present, composites are not used in clinical practice but there are some studies on the possibility to use these materials as vascular grafts. Porous polyethylene terephthalate (PET), polytetrafluoroethylene (PTFE) or polyurethane (PU) impregnated with collagen or gelatin have been studied. Graft impregnation allows the preclotting procedure to be avoided which is needed to minimize blood leakage; it improves the dimensional stability of the graft and seals the pores. After the collagen and gelatin in vivo degradation, the porous wall of the graft allows the ingrowth and firm anchorage of the host tissue to the material. An arterial prosthesis made of PU fibers and a mixture of PU and PELA (a copolymer of lactid acid and polyethylene glycol (PEG)) has been obtained by filament winding technology. At the time of implantation the graft is dense preventing any blood loss, but the matrix is resorbable in vivo, thus resulting in the formation of a porous wall during the healing process, that allows the tissue ingrowth into the pores. Fiber orientation, amount and properties are selected in order to produce elastic and mechanical properties that closely match the properties of the natural vessel.

Applications in Dentistry

The hard dental tissues, mainly enamel and dentine, are natural composites made of collagen and hydroxyapatite (HA) microcrystals. The use of composites materials in dentistry, mainly polymeric matrix composites, has considerably grown in the last years substituting in some cases more traditional materials. Polymer matrix composites are used in restorative dentistry to fill cavities, to restore fractured teeth and to replace missing teeth.

Dental Restorative Materials

Composite materials are used in clinical practice to restore anterior and posterior teeth. They have replaced polymethylmethacrylate (PMMA) fillings while amalgams are still used for posterior teeth because they exhibit better compressive mechanical properties. In most applications, dental composite consists of a polymeric acrylic or methacrylic matrix reinforced with ceramic particles. The commercial formulations of matrices are mainly based on bisphenol-A-glycidyldimethacrylate (Bis-GMA); triethylenglycoldimetachrylate (TEGDMA) is added to reduce the viscosity.

The most critical aspects for long-term performance of dental composites are the dimensional stability, following the polymerization shrinkage, wear resistance, and mechanical properties. Wear and mechanical properties strongly depend on the filler-matrix adhesion; microcavities, resulting from material detachment, could play a decisive role in the adhesion of bacterial plaque. Finite elements analyses have shown that the maximum stresses generated by shrinkage are in the region of the interfaces. The higher the interface stiffness, the lower the deformation of the restorative material, but the stresses increases.

The research is devoted to optimize the performance of the material and to overcome problems associated with the composites that are currently used. The monomer-polymer conversion greatly affects the mechanical and biological properties of dental composites and can be varied and controlled by changing the cure conditions. Low degrees of conversion during the resin polymerization are responsible for the loss of mechanical, chemical and physical properties. TEGDMA release has been mainly studied and the quantity released was sometimes higher than the cytotoxic limit. The cytotoxic effects became less important when the curing time of the resin was increased and with preincubation periods in a biological medium.

Wear is still a cause of clinical failure, particularly in high stress areas such as molar restorations. The wear rate determined by occlusal loading of some commercial dental composites has been tested with respect to the reinforcement particles dimensions. Wear rate for composites with small particles was higher than for those with large particles. The in vitro experimental results do not reflect exactly the in vivo behavior of the materials because the effects of feeding and patient's conditions are not considered. A composite material may perform satisfactorily in one patient while, in another patient, the same material may wear, degrade or fracture. The consumption of alcoholic beverages, for example, produces damages in the composite because the alcohol acts as a plasticiser of the matrix.

Most research efforts are devoted to overcoming the problems related to the polymerization shrinkage of the composite matrix. The polymerization kinetics appear to be the most important parameter contributing to the magnitude of the internal stresses. The stress distribution is nearly the same in self - and light-curing composites but the magnitude of the internal stresses doubles in lightcured composites. The addition of inorganic filler particles reduces matrix shrinkage, while diluent monomers with a molar volume lower than that of Bis-GMA, as TEGDMA, increase matrix shrinkage and water uptake causing decay of the mechanical properties. Bis-GMA dimensional stability and mechanical properties may be improved by using monomers with a greater molar volume. Low viscosity dimethacrylate monomers requiring less diluent monomers have been studied. Many research efforts are also directed to the development of non-contracting resins and fillers with expansion capacity, but the optimal material has not been found yet.

Adhesion between the matrix and the filler is a critical issue for material performance. Coupling agents improve bonding between composite constituents. Different coupling treatments have been analyzed to evaluate the reduction of contraction stresses caused by resin polymerization. The tested matrix was made by 25 wt% Bis-GMA, 35 wt% TEGDMA and 40 wt% urethandimethacrylate (UDMA); the reinforcement was unsilanated, nonfunctional silanated and functional silanated silica particles. Particles not treated with a functional silane agent do not bond to the matrix, providing internal sites for stress relief. Hardness and moisture sorption measurements can be used to monitor the properties of dental composites. Hardness is related to the stability of the resin-filler interface and may monitor the effectiveness of silanation process. A composite made of a blend of UDMA and ethylene-glycol dimethacrylate, reinforced with glass and silica particles, has been tested. The hardness increases with volume fraction of the filler and it is significantly higher for silanated filler than the equivalent unsilanated filler. UDMA, reinforced with silica-zirconia particles, has been tested to investigate the effect of moisture on the stability of the matrix-filler interface (16). The addition of a filler increases the moisture sorption. The presence of a silane coupling agent improves the material stability against the hydrolytic degradation. Filler particles of small dimension (<20 μm) lead to a better surface finish, but are less resistant to hydrolytic degradation.

The possibility of aligning the filler particles using an electric field has been analyzed to improve the mechanical properties of dental composites. The analyzed matrix consisted of 90% UDMA and 10% 1,6-hexanediol dimethacrylate (HDDMA), as diluent agent, and the filler was silanated silica-zirconia. The alignment of the reinforcement increases both the elastic modulus along the alignment direction and the maximum compressive stress.

Some studies are dedicated to the development of innovative materials to be used in posterior fillings and multiple-unit restorations. An antibacterial monomer (MDPB) has been incorporated in Bis-GMA based composites to improve durability and stability. The results have shown that the antibacterial effect did not decrease even after aging in a wet environment for a long period and the material did not release antibacterial components in the oral cavity. Improvements in the mechanical properties, mainly brittleness and toughness, have been achieved with an innovative fibrous reinforcement. The analyzed matrix consisted of 50 wt% Bis-GMA and 50 wt% TEGDMA, while the filler was silicon carbide or nitride whiskers. Silica particles were fused on the whiskers to facilitate silanisation, to minimize whiskers entanglement and to enhance whiskers retention in the matrix by providing roughness.

Silicon nitride whisker reinforcement confers a twofold improvement in the flexural strength of dental composites, offering, at the same time, a high elastic modulus and fracture toughness, and a low brittleness with respect to current dental materials. Silicon carbide whiskers doubled the fracture toughness with respect to the current dental composites, while the brittleness was reduced to one half. HA-Ca polycarboxylate composites are studied for dental restorations. The material has shown good adhesion to teeth wall. HA addition improves biocompatibility and mechanical properties. In situ composites made of porous apatite and 40 wt% PMMA have been studied for applications as dentine restoration materials. A reinforcement shaped as a three-dimensional fibrous scaffold has also been studied. The fibers are made of 25 wt% alumina and 75 wt% silica and the matrix is based on Bis-GMA and TEGDMA. The reinforcement is well incorporated into the matrix, reduces the polymerization shrinkage and improves the wear resistance. Further work is required to optimize the material production.

Composite posts made of unidirectional carbon fibers embedded in an epoxy-resin matrix have been proposed in. Minimizing the mismatch of the elastic modulus between the post and the dentine reduces the occurrence of root fractures and restoration failures. While traditional metallic posts are subject to corrosion and fatigue failures, composite posts meet the requirements of mechanical strength and corrosion resistance. An ideal post should have varying stiffness along its principal direction: higher stiffness in the upper section, for better retention and rigidity of the restoration, and lower stiffness, close to that of dentine, in the lowest section, to avoid dangerous stress concentration. A functional composite post, with a constant cross-sectional geometry and made of epoxy resin reinforced with unidirectional carbon fibers, at different concentration along its axis is under investigation.

Fixed and Removable Dental Prostheses

Fixed bridges and removable dental prostheses are used to replace one or more teeth, thus restoring the masticatory, phonetic and aesthetic functions. Their main disadvantages are, fixed bridges interfere with the adjacent teeth through the points of attachment, removable dentures are uncomfortable and do not transfer the occlusal stresses to the bone. Some composite materials are used as bridges (carbon, Kevlar, ultra high molecular weight polyethylene (UHMWPE) or glass fibers reinforced PMMA) and as removable prostheses (glass fibers reinforced polycarbonate (PC), polyamides, polypropylene (PP) or PMMA). Good results, in terms of mechanical properties and aesthetic appearance, have been obtained, for applications as bridges, combining two different composite materials made of a methacrylic matrix reinforced with glass fibers and inorganic particles.

Applications in Oral and Maxillofacial Surgery

The most important reasons for oral and maxillofacial surgery are, replacement of damaged or missing teeth, replacement of resorbed bone tissue, replacement or repair of temporomandibular joint and facial bones and fixation of fractured facial bones.

Dental Implants

Dental implants have been developed to overcome the problems associated with bridges and removable dental prostheses. The damaged or missing tooth is replaced by an artificial permanent implant. Research and clinical practice are devoted to reach a total bone regeneration and to inhibit both migration of epithelial cells to the implant and fibrous encapsulation.

Composite materials have been studied and developed mainly for application as endosseus element. Composites made of SiC and carbon or carbon fibers reinforced carbon have been proposed. These materials have some advantages over ceramics and metal alloys. They combine sufficient strength with a low elastic modulus, similar to that of natural teeth, minimizing the alteration of the stress field affecting the long-term success and host response. Moreover, they have far superior fatigue properties. Some studies have addressed the combination of Ti with HA to optimize and satisfy both mechanical requirements and biological response of the host tissue to the implant. Graded composites combining Ti and HA has been tested in. The graded structure in the longitudinal direction contains more Ti in the upper section and more HA in the lower section. As a matter of fact, in the upper section the occlusal force is directly applied and Ti offers the required mechanical performance; in the lower part, which is implanted inside the bone, the HA confers the bioactive and osteoconductive properties to the material. This composite has been tested in vivo, no inflammation has been observed and new bone has grown at the interface between the bone and the implant. Ti alloy substrate has been combined also with HA granules spread over the surface, but further studies are needed to evaluate the biofunctionality and biocompatibility of this material.

Other studies are devoted to guided tissue regeneration. A composite membrane consisting of polyethylene glycol terephthalate (PEGT) and polybutylene terephthalate (PBT) copolymer reinforced with 30 wt% HA grains has been proposed in. This composite has excellent mechanical properties and in vivo experimental results reveal good biocompatibility and a gradual degradation, mainly influenced by the PEG/PBT ratio. The experimental results are promising but more in vivo testing and clinical studies have to be performed before clinical application.

Temporomandibular Joint Prostheses

Maxillofacial surgery is necessary for patients afflicted with arthritis of the temporomandibular joint. Wear is one of the most important reasons for long-term failure of temporomandibular joint prostheses. At present, the composite materials are not used in clinical practice. A composite cup made of PTFE and carbon or alumina has been tested as an alternative to UHMWPE cup, but the clinical results have shown significant inflammatory response near the implant site. Further studies are needed to develop the optimal material and prosthesis geometry.

Cranial Bones Repair

Fracture fixation devices are employed to guarantee rigidity during the healing process. Intense research activity has been devoted to the use of bioactive ceramics, mainly TCP and HA, because they have composition and morphology close to those of the mineral phase of natural bones. These osteoconductive materials promote and assist the healing of the bone defect, by acting as a temporary scaffold for bone ingrowth. A biodegradable and osteoconductive composite made of tricalciumphosphate (TCP) particles and gelatin has been proposed in. This material has been tested in vivo with good results. It has shown itself to be biocompatible, osteoconductive and biodegradable, with no need for a second surgical operation to remove the device after healing has occurred. Further research is needed to obtain the repair of large defects. Miniscrews and miniplates made of poly-L-lactid acid (PLLA) matrix reinforced by u-HA (raw hydroxyapatite neither calcined nor sintered) particles, respectively 30 wt% and 40 wt%, have been manufactured and tested. They have shown easy handling and shaping according to the implant site geometry, total resorbability, good ability to bond directly to the bone tissue without interposed fibrous tissue, osteoconductivity, biocompatibility and high stiffness retainable for the period of time necessary to achieve bone union.

Applications in Tissue Engineering

Tissue engineering is an emerging area in medicine involving cells manipulation to promote the regeneration and healing of defective natural tissues or the development of biological substitutes that restore, maintain or improve tissue and organ functions. The combination of natural or synthetic scaffolds with cells may provide an alternative to both organ transplantation and implantation of mechanical medical devices unable to perform all functions of a tissue or an organ.

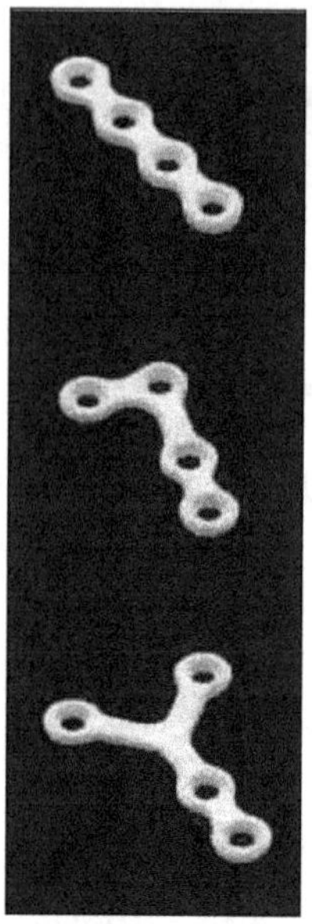

Miniplates made of 40 wt% u-HA reinforced PLLA.

Some composite materials have been studied to verify their performance in applications as scaffold for cellular growth in tissue engineering. Some studies are devoted to the development of a scaffold which combines specific biological properties and electrical characteristics, improving bone and nervous tissue regeneration. Composites made of hyaluronic acid and polypirrone have been synthesized and characterized. The composite has been tested in vitro and in vivo showing good results. For tissue engineering of bone and cartilage, a composite consisting of a three-dimensional mesh composed of 80 wt% polycaprolactone (PCL) fibers with 20 wt% HA particles on the surface has been evaluated. The ideal scaffold material has not been developed yet and further investigations are needed.

Other studies deal with the combination of PCL and HA with different morphology. The composite is obtained by infiltration of ε-caprolactone monomer into porous apatite blocks and in situ polymerization. The material is biodegradable and may have application as cancellous or trabecular bone replacement material or for cartilage regeneration. The polymer molecular weight increases with the polymerization time, leading to an improved compressive strength.

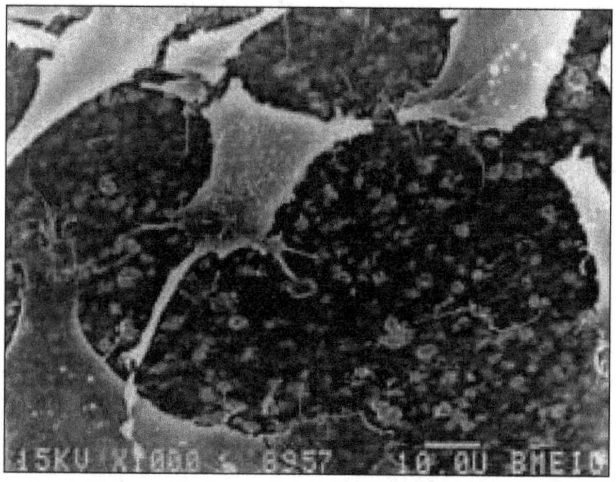

Osteoblasts adhesion on HAPEX surface.

Orthopedic Applications

Orthopedics is the medical area where application of the biomaterials is highly developed both for commercialization and research purpose. The use of composites in orthopedics offers a variety of new implant design and the possibility to tailor specific device properties. Composite materials can reproduce both the macroscopic and microscopic structure and the most important mechanical properties of the natural tissues. Composites materials are studied and tested to improve the performance and the long-term stability of femoral stems and bone cement, to replace cartilage, to reconstruct tendons and ligaments and as bone grafts.

Bone Graft

The bone tissue combines a cellular component, the osteocytes, and a matrix. The matrix is a composite made of an organic portion, composed of collagen fibers, giving the bone strength and flexibility, and an inorganic portion, composed of HA microcrystals and mineral salts, conferring hardness to the bone. The bone tissue is continuously replaced and remodelled according to the applied stresses by the osteoblasts and the osteoclasts, and it is able to self-regenerate after injuries. However, when injury is too severe and loss of volume is too high, healing is incomplete, fibrous tissue forms, mechanical functionality is not restored and bone grafts are necessary. The ideal bone graft is represented by autografts, but their amount is limited, their properties and shape do not match exactly those of the bone to be replaced, and multiple surgical operations are needed. The use of allograft is accompanied by biological and mechanical difficulties and the risk of infection transmissions exists; for these reasons synthetic bone grafts are under extensive investigation. Some studies are devoted to the development of a composite graft consisting of demineralised bone powder packed between two collagen layers. The laminate has been tested in vitro and in vivo showing cell migration through the collagen layers and the interaction with the bone demineralised powder.

Great attention is payed to the use of bioactive ceramics because of their ability to bind to the natural bone tissue. The contact with blood plasma induces the surface coating with a bone-like HA layer where the osteoblasts can proliferate. Most studies deal with the combination of high density polyethylene (HDPE) and HA particles as bone replacement materials. The resulting composite has been tested in vitro, in vivo and clinically evaluated. The successful results have allowed its commercialization in 1995 with the trade name of HAPEXTM. The effects of the reinforcement content, the size of the HA particles and the addition of coupling agents on the biological and mechanical properties have been analyzed. The range of HA fraction has been optimized between 20 and 40 vol%, giving a bioactive and fracture tough implant: HA fractions less than 20 vol% caused the material to become inert promoting a fibrous encapsulation of the implant and a ductile-brittle fracture transition exists at about 40 vol%.

Composites with small HA particles show a high torsional modulus, tensional modulus and tensile strength, but a low strain to failure. HAPEX stiffness and strength are below the values associated with cortical bone; consequently, this bone substitute cannot be used in major load-bearing applications. The experimental results have shown a good adhesion and proliferation of osteoblasts on HA surface particles. Increasing with the HA content. The mechanical performance depends on the filler-matrix interface which affects the long-term success and stability in the biological environment. The examination of the fracture surfaces revealed that only mechanical bond occurs between HA and polyethylene (PE). To improve the mechanical performance of HAPEX, it is necessary to introduce a coupling agent at the filler-matrix interface for an effective stress transfer and to promote chemical bonding between the reinforcement and the matrix. Further improvement could be obtained with the molecular orientation in the polymer matrix and the addition of woven, high performance PE fibers. Many studies deal with the development of a bioactive material for bone replacement with composition and biological response similar to that of natural bone.

Composites obtained by the combination of HA particles and collagen fibers have been analyzed in. The graft has revealed a very high biocompatibility, excellent biological properties and good mechanical performance. HA/alumina composites have been proposed as bone replacement materials. In vivo results have shown that no soft tissue is present, the composite bone-bonding is a function of the content of HA, while the mechanical properties are connected to the alumina content. P_2O_5 glass reinforced HA composites have been developed to synthesize a bioactive graft material with a stiffness close to that of a natural bone. The elastic properties are strongly dependent on the material porosity. Some studies are devoted to the development of partially resorbable composites made of a polymeric matrix, such as PEG, PBT, PLLA, polyhydroxybutyrate (PHB), alginate and gelatin, and a bioactive filler. The material achieves a good integration into the bone since the matrix resorption yields a large number of bioactive particles coming in contact with the tissues. A coupling agent was added to the composite made of PEG/PBT copolymer (ratio 70/30) matrix and HA reinforcement improving the mechanical properties by covalent bonding of the HA with the polymer matrix.

Elastomeric D, L-lactide and ε-caprolactone copolymers (ratio 60/40) reinforced with HA powder was analyzed to characterize the effect of the HA content. The incorporation of HA reduces the degradation rate and has effect on the mechanical and handling properties. HA reinforcement morphology different from particles, such as HA fibers, have been studied and characterized. Biodegradable composite foams of PLGA, a copolymer of polyglycolic acid (PGA) and poly lactid acid (PLA), (PLA/PGA ratio 85/15) reinforced with HA short fibers have been tested for use in bone regeneration. The matrix porosity facilitates the migration of osteoblasts from the surrounding bone to the implant site. The HA fibers enhance material mechanical properties and promote the bone regeneration. Further studies are needed for the optimization of the composite graft porosity to allow cell seeding without loss of mechanical properties.

Among the bioactive ceramics, TCP, as well as HA, stimulates great interest for application as bone graft. A bone replacement composite made of α-TCP reinforced with polyammide (PA) fibers has been proposed in. The addition of fibers produces a moderate increase of the compression strength because of the weak coupling of the fibers with the matrix. Higher compression strengths can be obtained only through modification of fiber surface or by using coupling agents. Composites constituted by TCP and glutaraldehyde cross-linked gelatin have been analyzed and tested in vitro to measure the material cytotoxicity. The experimental results have shown that the glutaraldehyde cross-linking agent added into the composite should be lower than 8% to decrease the toxicity on the osteoblasts and to avoid the inhibition of cellular growth caused by the release of residual or uncross-linked glutaraldehyde. TCP has been also combined with biodegradable PP fumarate to verify the possible application as temporary replacement for trabecular bone. In vivo results proved that the composite is biocompatible and allows a progressive new bone growth inside the degrading material.

The composite maintains a compressive strength and modulus greater than the minimum value for trabecular bone replacement. In addition to HA and TCP, bioglasses are also studied as bone replacement materials. Composite materials have been prepared using 37 wt% bioactive glass powder, 27 wt% PLA, 27 wt% PMMA and 9 wt% antibiotic, gradually released to prevent infections and inflammations. In vitro experimental results have shown the growth of an apatite-like layer on the composite surface. Further investigation is needed to evaluate the long-term effects of the degradation products, and to verify whether the growth of a surface apatite layer inhibits the antibiotic release. Bioglass reinforcements have been tested also in combination with a PE matrix. The matrix/reinforcement adhesion is weak because there is only mechanical coupling. The composite has shown mechanical properties very close to that of natural spongy bones with bioglass content higher than 30 vol%. The adhesion between the matrix and the reinforcement is weak, if compared to HA/PE composites, and the elastic modulus is low, but the osteoconductive properties are high and this composite needs short times to be integrated with the bone tissue surrounding the implant site.

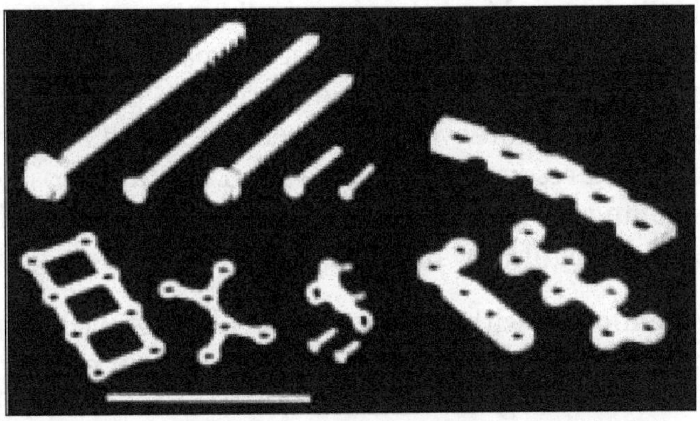

Internal fixation devices made of u-HA reinforced PLLA.

Carbon fibers-reinforced carbon has also been tested in vivo. The results have shown that a mechanical bonding between the implant and the host tissue does exist, mainly due to the initial porosity of the material, and there was no connective tissue encapsulation around the implants. A composite implant of a coral (mainly $CaCO_3$) and collagen has been tested in vivo for use in the healing of bone defects. There was no infection after implantation and the implant is well integrated in the natural tissue with no interposed fibrous tissue.

Bone Fracture Internal Fixation Devices

Internal fixation devices are temporarily implanted inside the body to hold together the bone fragments and promote healing. The critical issues associated with the use of metallic internal osteosynthesis devices are, metals are exposed to corrosion in the physiological environment and the stiffness mismatch between the bone and the metallic implants induces the stress-shielding phenomenon affecting bone remodeling and healing process. The research spent in this field is justified with the assumption that an ideal internal fixation device should have a high mechanical strength to avoid fracture, but a stiffness close to that of the natural bone tissue.

Polymeric composite materials may offer the desired high strength and bone-like elastic properties. Composite fixation devices based on a thermosetting matrix may induce toxic reactions following the release of unreacted monomers. Thus, the research activity is oriented to the development of thermoplastic composites, such as carbon fibers reinforced PMMA, polysulfone, PP, PE, PBT or polyetheretherketone (PEEK). Mechanical and biocompatibility tests have shown good results, but there are not yet enough data on long-term in vivo performance.

A novel composite made of unidirectional carbon fibers, embedded in a matrix of liquid crystalline polymer, has been proposed in. This material has been tested in vitro and in vivo showing a flexural strength higher than that of some metals used in orthopedic applications, an elastic modulus quite close to the corresponding value of cortical bone and a good chemical resistance. The initial mechanical properties are retained after implantation.

The use of composite materials made of a polymeric biodegradable matrix is an interesting research field. Their main advantages over metals are no second surgical operation is necessary, the fractured bone has a low tendency to refracture after the removal of the implant, the device may be engineered to degrade at a rate allowing a gradual transfer of the load from the implant to the healing bone. Furthermore, there are no tissue reactions caused by metallic corrosion and drugs can be inserted and gradually released into the surrounding tissues. Totally bioresorbable internal fixation devices made of PLLA reinforced with raw u-HA particles have been manufactured and characterized. The composite has ultra-high strength, excellent processability and an elastic modulus close to that of natural bone. Moreover, it can retain its high strength during bone healing, shows optimal degradation and resorption behavior, and maintains osteoconductivity and bone bonding capability. This composite has been tested in vitro, to analyze the degradation process, and in vivo, to verify the biocompatibility and the bioactivity, showing good results.

Composites made of PLA reinforced with HA fibers, with almost stoichiometric composition, have been studied in. The composite is able to consume a large amount of energy in the fracture. Polymeric composites reinforced with bioglass particles have been studied and tested in vitro to verify the bioactivity and osteoconductivity after immersion in physiological solution. PLA composite rods reinforced with 15-25 wt% bioglass particles, PLA plates coated with bioglass spheres, with a total glass content of 17.5 wt%, and polyethyleneoxide (PEO)/PBT copolymer rods reinforced with up to 40 wt% bioglass particles have been analyzed. All these materials show a good bioactivity leading to the formation of an appetite layer on the surface. Bioglass reinforced PEO/PBT has shown the best and fastest apatite formation.

A critical issue of biodegradable matrices is the degradation rate control during fractured bone healing. The loss of device mechanical properties during the degradation should be synchronized with the healing rate. At present, the non-optimal mechanical properties limit the application of biodegradable composites to internal fixation devices used in reduced load conditions. Nevertheless, the future perspectives are promising.

Joint Prostheses

The use of composite materials in artificial joints is a promising research field. The implant properties can be varied and tailored to suit the correct mechanical and physiological response. Prostheses with stress and strain distribution close to physiological ones have more chances of long-term stability and clinical success. The current results are promising, but more experience is needed regarding the clinical use of composite artificial joints. As a matter of fact, to carry out clinical experiments extended periods of time are needed and some concerns persist on the improvement in biocompatibility using composite materials.

Hip

The hip prosthesis is the most implanted artificial joint. Clinical data show that about 90% of artificial hip joints are successful for the first 10 years. Eventual revision surgery reduces the long-term success. The main failure reasons of total hip replacements are the stress shielding phenomenon, the relative movement of the prosthesis, leading to the development of a soft tissue interface between the bone tissue and the prosthesis, and the presence of wear particles that may induce adverse biological reactions in the body. Moreover, metallic components undergo corrosion phenomena in the physiological environment, which may induce stem fracture and cytotoxic and allergenic processes.

Composite materials are not currently used in clinical practice for total hip replacement, but they are studied with great interest for application as acetabular cup and femoral stem. Acetabular cups are made of ultra-high molecular weight polyethylene (UHMWPE) and are still the weakest link in a hip prosthesis because of UHMWPE wear, erosion and creep deformation. To increase their long-term performance, many research efforts are devoted both to improve the mechanical properties of UHMWPE, by adding reinforcements, and to develop new alternative materials. Acetabular cups made of UHMWPE, reinforced by carbon fibers or UHMWPE fibers have been investigated, but the effect of the fibers on wear behavior are not clear yet, and the reported results are contradictory.

A carbon reinforced PEEK has been proposed as an alternative for UHMWPE. Experimental results have shown that the wear rate decreases with high fiber content, and that the best option for good wear characteristics and dimensional stability is an acetabular insert made of 30 wt% pitch-based carbon fibers reinforced PEEK and a zirconia head. With this combination, the wear rate has been reduced by almost two orders of magnitude, compared to a conventional UHMWPE/metal or UHMWPE/ceramic combination. Carbon fibers reinforced epoxy resin has been analyzed as an orthopedic bearing surface for acetabular cups in. The combination of alumina femoral heads and composite cups results in excellent long-term behavior minimizing the wear debris and biological response. The wear rate is respectively 5 and 30 times lower than UHMWPE cups coupled with alumina and UHMWPE cups coupled with metal.

Composite materials have been investigated also as coatings or as the main constituent for femoral stems. Bioactive composite coatings have been obtained combining a bioglass and a titanium alloy (Ti_6Al_4V). The bioglass gives osteoconductive properties to the coating, while Ti makes it tougher. For metallic prostheses with a bioactive coating there is still a mismatch between the natural bone tissue and implant mechanical properties. Composite femoral stems have been developed to reduce the problems associated with the stress shielding and the micromotion. Stems with a controlled stiffness, close to that of natural bone, can be obtained by combining a stiff material with a more flexible one and varying the thickness of layers.

A femoral stem made of a metallic cobalt-chrome core surrounded by a composite layer has been analyzed in. The composite layer was made of an epoxy resin reinforced with carbon and glass fibers. The studies on femoral stems entirely made of composites have shown good results. The strength of composite prosthetic components may be changed without affecting the stiffness and the modulus can be varied without altering the strength. Finite elements analyses reveal that composite stems allow a more uniform and effective stress transfer, close to the physiological one, than metallic stems.

Nevertheless, there are two limiting factors: given an applied load, higher strains are generated inside more flexible implants and the shear stresses at the interface between the implant and the bone is high for those implants with a low stiffness. Carbon fibers reinforced PEEK is one of the most promising composites considered for the femoral component. In vitro aging results have not shown a significant loss of compressive strength and modulus. In vivo results have shown mechanical properties very close to that of natural femurs, minimal tissue reaction and a mechanical stability of about 6 months. Polysulfone and carbon reinforced with carbon fibers have been also proposed for femoral stems. The experimental results have shown fast bone bonding, when compared to metallic prostheses, a good stability without the release of soluble elements and a high static and fatigue resistance. Some studies are devoted to the analysis of laminate composites because the loads applied to the hip joint are not unidirectional.

A composite stem made of 120 epoxy resin laminae, reinforced with carbon fibers, stacked according to a specific sequence has been tested. Thermosetting polymers may release toxic unreacted monomers in the surrounding tissues. Thus, thermoplastic polymers are preferred as composite matrices. The use of polyetherimide (PEI) is very promising due to its excellent biocompatibility. PEI reinforced with carbon and glass fibers have been tested in vivo. Composite implants do not induce foreign body reactions and their response is comparable to that of metallic devices. A composite laminate femoral stem made of a sequence of laminae of PEI reinforced with carbon and glass fibers has been manufactured and characterized. The laminate prosthesis is characterized by a stiffness similar to that of the surrounding bone, ensuring a load transfer between the implant and the bone similar to the physiological one, and by a high strength, preventing the implant from fracture.

Knee

The knee joint is the most exposed to traumatic and degenerative injuries. Knee geometry and kinematics are more complex than those of hip, even if with lower loads. This is why different geometries and types of knee joint replacements have been developed and the long-term success of the implants is lower than for hip replacements. At present, composite materials are not used in clinical practice. They have been tested for the tibial component to overcome the problems associated with UHMWPE creep and wear. Creep deformation induces sinking of the prosthesis while wear particles induce a foreign body reaction.

However, the results are contradictory and not clear enough to foresee the long-term material performance. A composite tibial component made of UHMWPE and UHMWPE or carbon fibers has been proposed. The reinforcement addition reduces the stress concentration during the articular movement and improves the mechanical properties. The results concerning the wear resistance with carbon fibers reinforcement are contradictory. Hence, some studies are oriented to combine a UHMWPE surface with a composite substrate so that the wear resistance would not be affected. No significant difference in the wear performance has been observed with UHMWPE fibers.

Bone Cement

The bone cement is a grouting material used to ensure a firm fixation of the joint prostheses. The most widely used commercial formulations are based on PMMA, are self-polymerizing and have minimal adhesive properties. Thus the bonding with the bone and the implant surface is a mechanical one consequent to the polymer setting and hardening. The critical issues of commercial bone cements based on PMMA are the exothermic nature of polymerization, that may locally induce bone necrosis, the shrinkage associated with the reaction, that may lead to mechanical failure and instability associated with the deterioration of the cement/implant or cement/bone interfaces, and the release of cytotoxic unreacted monomers and additives in the surrounding tissues and in the blood stream.

Much effort has been made to improve the mechanical properties of bone cement and to overcome the key problems. The first attempts tried to combine PMMA with fibrous reinforcements, such as carbon, Kevlar, stainless steel, Ti, glass or PE fibers. The addition of fibers resulted in big improvements in the mechanical performance, but fibrous cements have not been accepted in clinical practice because of a loss of material workability, consequent to the viscosity increase, and the difficulty of obtaining a uniform distribution of the reinforcement inside the matrix. Better results have been obtained with particles reinforced cements. The performance of partially resorbable bone cement made of a gelatin reinforced with TCP particles has been evaluated. This composite allows the bone ingrowth and the direct chemical bonding between the TCP particles and the bone tissue following the matrix resorption.

On the other hand, the mechanical properties of partially resorbable bone cements are lower than that of PMMA cement. The introduction of a bioactive phase may enhance the prosthesis fixation by encouraging the direct bone apposition rather than the encapsulation of the implant by the fibrous tissue. Bone cements made of PMMA and bone or bioglass particles have been proposed. The mechanical properties and the bioactivity of cements constituted by PMMA and HA have been also investigated. The addition of HA allows the direct bone apposition and offers a better substrate for osteoblasts attachment and proliferation, compared to a PMMA cement. As a drawback, the PMMA/HA cement has shown a low flexural, compressive and tensile strength. Further investigations are needed to optimize the PMMA/HA ratio in order to obtain improved mechanical performance. A composite cement based on PMMA and HA particles has been compared to a cement made of PMMA and bioglass particles (30 wt%). When implanted in vivo, PMMA disturbed the osteoconductive bone formation on the surface of bioactive materials. This effect was less evident with bioglass than with HA.

New composite cements with matrix different from PMMA have been developed. Composites with polyethylenmethacrylate (PEMA) based matrix and HA reinforcement cement have been tested. The presence of HA increases the matrix fatigue and flexural properties and reduces the water uptake responsible for the release of unreacted monomers and polymerization residues, the breakdown of the polymer itself, the penetration of microorganisms and the decrease of mechanical properties. In vitro experimental results have shown that this composite cement produces lower temperature at bone/cement interface and gives higher conversion rates than commercial PMMA cements. A composite cement based on Bis-GMA and 70 wt% of alumina powder has been tested in vivo, showing an excellent osteoconductivity and a high mechanical strength. The alumina fillers were surrounded by bone-like tissue and no interposed soft tissue was observed.

The ideal bone cement has not been developed yet. Among its characteristics, this material has to be easy workable and should allow good adhesion to the bone by penetrating into the internal cavities of the bone tissue. Also, the mechanical properties have to be adequate to the implant stabilization and should not prevent the bone remodelling, offering a low reaction temperature and a high dimensional stability.

Artificial Tendons

The tendons can be considered as unidirectional fibers reinforced composites made of collagen fibers, aligned along the principal direction and immersed in an elastine-mucopolysaccharide hydrogel. When stress is applied the collagen fibers line up along the stress direction; if the stress is higher than the fibers strength resistance, a fracture occurs. Hence, the mechanical behavior depends on the matrix properties at low stress values and on fibers properties at high stress values. The tendon tissue allows a large extensibility without permanent deformations. Most of the commercial tendons prostheses undergo mechanical failure due to an inadequate fatigue, especially bending and friction, resistance.

Great attention is given to the design and fabrication of composite tendon prostheses reproducing the structure, resistance and flexibility of the natural tissue. Most of the artificial tendons studied until now are based on polyesters and hydrogels. The mechanical behavior of a composite made of poly2-hydroxyethylmethacrylate (PHEMA) matrix and PET fibers reinforcement has been studied in. The composite reproduces the stress-strain curve showed by the natural tendons under low strains. A different mechanical response can be obtained varying the alignment of the fibers. A composite made of ethylene-butene copolymer and UHMWPE has been tested in, to characterize the fatigue resistance as a function of the copolymer composition and the fibers alignment. Good long-term performance is ensured by more branched copolymers and high angles between the reinforcement and the applied load direction. A composite material based on polyester and chitin has been analyzed to evaluate chitin effects. The chitin is a wound healing accelerator and, in vivo, it induces collagen and elastic fibers in the prostheses. Further investigation is needed to clarify this mechanism.

The surgical and long-term implant success depends on the implanted materials, but also on the firm fixation of the prosthetic element to the bone. The biological attachment of autogenous tendons to the bone tissue has been studied in vivo by injection of a composite made of 40 wt% collagen and 60 wt% powdered HA, between the grafted tendon and the bone tunnel. The collagen was gradually degraded and replaced by the regenerating bone in direct contact with the grafted tendon without an interposed fibrous tissue layer. The collagen fibers of the grafted tendons penetrate the newly formed bone, increasing the implant stability.

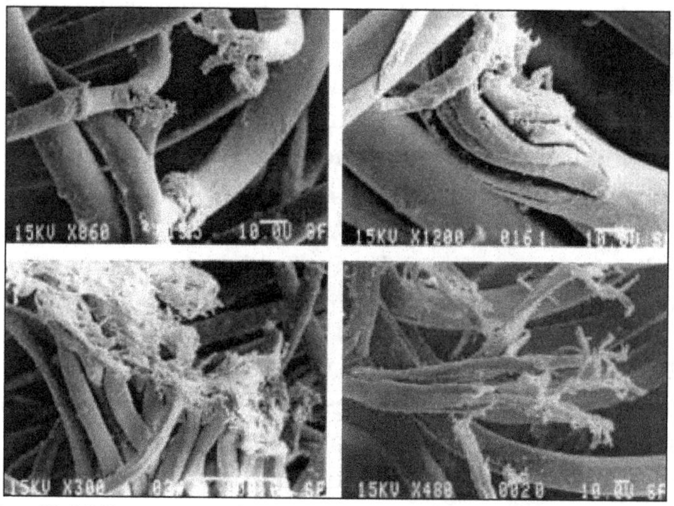

Commercial artificial ligament explanted because of fatigue failure and synovitis onset.

Further investigations are needed to discover the ideal material for tendons replacement, which might ensure long-term performance and stability. The complexity is increased by the variation of the geometry and properties of a tendon during growth and development, mainly the cross-sectional area, modulus and strength, as a function of the mechanical loading.

Artificial Ligaments

The ligaments have a complex fibrous anatomy and a non-uniform structure and, as for tendons, their geometry and properties vary during growth. Most ligament injuries are related to the knee anterior cruciate ligament (ACL). Commercial artificial ligaments are polymeric or polymer-composite devices. The success rate ranges only between 30 and 60%. Mechanical failures are mainly due to the weak fatigue resistance, especially in the aggressive human environment and under cyclic loads. The inadequate fiber resistance to abrasion and flexural and torsional fatigue, and synovitis are among the main reasons of long-term failures of artificial ligament.

A biodegradable prosthesis, made of PLLA and ialuronic acid esther, has been studied. The gradual degradation allows the replacement of natural tissue. Some composites ana lyzed for tendons replacement have been tested also for artificial ligaments applications, as the UHMWPE reinforced ethylene-butene copolymer and the PET reinforced PHEMA. Experimental results show that PET may induce synovitis. Thus, terephtalate polyester woven fibers have been combined with an extracellular matrix based on collagen. The prostheses have been tested in vivo showing no fracture, no signs of inflammation or foreign body reaction and the growth of oriented fibrous tissue between the bone and the prosthesis. Artificial ligaments are being progressively dismissed, leading to a more frequent use of autografts from rotulian tendons. Hence, in many cases, the prosthesis has the function to support the autograft (augmentation devices).

Artificial Cartilage

Cartilage replacement materials are used to repair congenital, developmental or acquired defects. A composite constituted by PET fibers reinforced PHEMA has been proposed to replace intervertebral discs. It is possible to design a material with mechanical properties comparable to those of natural discs by varying the fiber content and the matrix composition. Further improvements for long-term performance are required. An innovative biocomposite made of a single UHMWPE long fiber with a three-dimensional structure, coated with u-HA microparticles, has been developed for potential applications as articular cartilage, menisci and intervertebral discs. When implanted, the surrounding tissues fill the textural space of this structure as a matrix, consequently forming a composite.

The implant becomes a part of the natural tissue as a consequence of mechanical and physiological integration with the natural tissue. u-HA particles confer surface bioactivity, allowing enhanced interactions with the surrounding natural tissues. The biocomposite has been tested in vivo showing no fiber wear, no infection, good biocompatibility and mechanical properties, but a reliable method to repair large defects has not been found yet.

Tissue Engineering

The concept of tissue engineering aims to regenerate damaged tissues with the support of scaffold biomaterials that act as a template to support the growth of new cells. Biocompatible scaffolds based on the natural extracellular matrix (ECM) are cell- and tissue-specific materials that aid the fast integration of tissues.

Human cells can attach and organize well around polymer composites with dimensions smaller than those of the cells. Polymer composites display a significant morphology which facilitates cells to migrate, grow, and seed. The improvement of polymer composites for cell proliferation and adhesion is important for organ and tissue reconstruction. In this concept, 3D biocompatible composites are required for cell growth for tissue repair and replacement. In general, polymer-based composites are promising templates to mimic the native morphology. To illustrate this, researcher processed porous biocompatible protein polymer thin films for use in implantable devices. Additionally, elastin-mimetic peptide polymers have been prepared and electrospun by researcher.

Bone

The bone of the skeletal system is considered as a morphological composite, consisting of collagen fibers with hydroxyapatite (HA) nanocrystals deposited along the collagen fibers. Bone also consists of other materials such as bone cells, blood vessels, and mucopolysaccharides. The collagen fibers exhibit low elastic modulus and they are aligned in bone along the main stress direction. The HA mineral covers 70% of the dry bone weight and attributes to the bone stiffness. It is noteworthy to mention that bone can adapt and remodel itself in response to the applied mechanical surrounding, which is well known as Wolff's law.

Bones are normally feeble with regard to shear and tension, mainly along the longitudinal plane. Fracture is one of the most common bone disorders that need special medical care. There are different kinds of fractures, depending on the location, structure, orientation, and crack volume. Proper implant systems and surgical procedures must be controlled to attain the required biomechanical fixation and to bypass any extra tissue devascularization and trauma at the fracture point. Moreover, the fracture healing process depends on the patient's activities, as they control the stability of mechanical performance at the fracture site. All implants are temporarily sited inside the patient's body and may be eliminated from the body after fracture healing.

Scientists have been paying special effort to tissue engineering to improve biodegradable bone graft substitutes. Different degradable polymers, either synthetic or natural, have been used as scaffold systems for bone tissue engineering, such as chitosan, poly(propylene fumarate), and polyesters, including the copolymer poly(lactide-co-glycolide) (PLAGA). For instance, PLAGA/calcium phosphate composite microsphere-based scaffolds were prepared by researcher via in situ formation of calcium phosphate within forming microspheres. The fabricated materials were incubated in simulated body fluid (SBF) for 8 weeks. It was hypothesized that a 11–14% drop in the composite scaffold molecular weight would occur, compared to a 20% drop in polymeric scaffolds over 8 weeks.

In addition, SBF pH and composite scaffold mass declined first and then increased after two weeks, assumed to be a result of the chemistry between calcium phosphate dissolution and pH variations. The free calcium ion content of SBF containing composite scaffolds surged from 20 to 40% within 4 h of incubation and then declined to 40%, contributing to the burst release of calcium ions followed by reprecipitation on the scaffold surface. Moreover, these composite scaffolds initiated calcium phosphate reprecipitation for an bone/implant integration approach.

Polylactic acid (PLA) exhibits poor tensile characteristics for weight-bearing fixation systems. Hence, reinforcements including glass, ceramics, and alloys have been blended with PLA chains to enhance the mechanical performance. Polymer composites containing phosphate glass fiber (PGF) displayed good biological and mechanical characteristics for repairing bone fractures. For the approach of the design of bone plates applied in long-bone fractures, researcher synthesized unidirectional bioglass fibers (BGF) (13–93)/PLA composites treated with air plasma. Different doses of atmospheric air plasma exposure (30, 60, 90, and 120 s) were used on the surface of BGF to develop the interaction properties with the PLA galleries. The fatigue life of 30 s plasma-treated blends achieved 1 million cycles by using actual content conditions of 10–20% body weight. After 30 s plasma exposure, it was found that the interlaminar shear strength, flexural strength, and tensile strength were surged by 33%, 13.5%, and 31%, respectively. Also, the superior interaction between PLA network and fibers was evidenced by the failure shifting to the PLA. More interestingly, bone-like calcium phosphate layers were deposited on the surface of the decomposed blends, which is required for bone healing processes.

In addition, various biometallic materials have been remarkably applied in internal fixation processes due to their strength and toughness. The desired implant should have the ability to degrade spontaneously inside the body to bypass its removal and stress shielding. For example, low-molecular-weight PLA was reinforced with various Mg rods by researcher to fabricate biodegradable PLA composite rods using a plastic injection molding (PIM) technique for bone fracture fixation. The researchers selected magnesium fluoride (MgF_2) in their experiments to strengthen the fabricated rods. In addition, hydrofluoric acid (HF) was used to produce a porous MgF_2 ceramic layer on the Mg rod surface to enhance the interaction between the internal Mg rod and PLA network. It was realized that PLA-clad Mg composites with an intermediate coating displayed much better bending strength, tensile strength, and corrosion resistance in simulated body fluid (SBF) solution compared to PLA-clad Mg composites without coating. In contrast, the bending and tensile strength were diminished clearly with the presence of the intermediate coating, owing to the corrosion of the MgO porous interface.

One of the most preferred types of material for repairing bone defects is the osteoconductive materials (such as HA), which could simplify the growth of osteoprogenitor cells. Recently, PLA/ethyl cellulose (EC)/HA composite scaffolds have been investigated as weight-bearing substitutes. The mechanical performance of these scaffolds was improved by combining the particulate leaching, high content solvent casting, and room temperature compression molding methods. The mechanical and hydrophilicity characteristics of the composites were improved after incorporating HA and modifying the scaffold surface by sodium trimetaphosphate (STMP).

The results revealed the PLA/EC/HA composites at 20 wt % HA content provided an excellent porous morphology and ideal mechanical behavior. The weight loss, compressive yield strength, contact angle, and porosity after 8 weeks were 4.77 ± 0.32%, 1.57 ± 0.09 MPa, 45.13 ± 2.40°, and 84.28 ± 7.04%, respectively. During hydrolysis, the scaffolds exhibited superior dimensional stability and porous morphology as well. In addition, alginate (AL)/HA/silk fibroin (SF) composites were fabricated by researcher as bone tissue scaffolds for rat calvarial defects. Figure represents the calvarial defects grafted with AL, AL/HA, and AL/HA/SF beads. To fulfill the criteria for a unique osteoconductive candidate for bone defect repair, polypropylene carbonate (PPC)/poly(d-lactic acid) (PDLA)/β-tricalcium phosphate (β-TCP) (PDT) malleable composites were designed by researcher. The tissue biodegradation, compatibility, and osteoconductivity of the composites were evaluated at up to 12 weeks after incorporation in rabbit femur bone defects. The bone defect was regenerated with phosphate-buffered saline (PBS) and PPC/PDLA/β-TCP (PDT) scaffold in six rabbits, while it was left un-regenerated in four rabbits. The findings revealed that the PPC/PDLA/TCP scaffold with a weight ratio of 90/8/2 was a biodegradable, biocompatible, osteoconductive, and malleable biomaterial for bone defect repair.

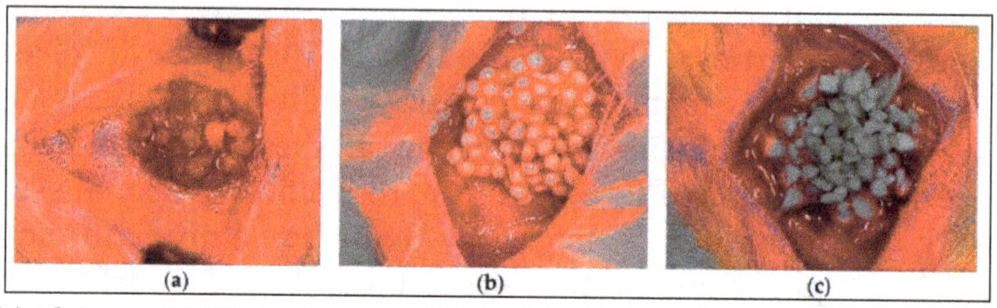

(a) Alginate; (b) HA/alginate; and (c) SF/HA/alginate beads were grafted into rat calvarial defects (silk fibroin (SF) and hydroxyapatite (HA)).

Blood Vessels

The repair of blood vessels is a challenging subject to researchers because most of the artificial scaffolds are not desirable for application, which leads to an increase of the time needed for patient recovery. One of the main biomedical tasks for patients who require peripheral vascular bypass surgery and coronary artery is the vascular regeneration. It is worth mentioning that tissue engineering vascular grafts, which combine synthetic vascular grafts with patient's cells to replace and repair the injured vessels, are used as one of the most promising potential templates for biomedical applications at present. To mimic the function and morphology of the extracellular matrix, electrospinning has been widely applied as a simple technique to fabricate 2D and 3D fibers using natural and artificial polymers. This technique was widely used to process tubular scaffolds of various diameters and lengths for vascular grafts. Electrospun tubular scaffolds with small diameters were processed from various synthetic polymers with vascular proteins.

Bacterial nanocellulose (BNC), structured by repeated β-1,4 linked D-glucose dimers, has been reported to provide some novel properties, such as an ultrafine nanofiber matrix; high water holding capacity, crystallinity, and chemical purity; and excellent biocompatibility and wet tensile strength. For instance, researcher evaluated BNC composite tubes doped with PVA as artificial blood vessels. The neat BNC tubes were considered to be an inadequate model for vascular grafts due to water leakage and poor suture retention. The water permeability and tensile characteristics were enhanced after blending BNC tubes with PVA. The mechanical properties of the fabricated composites were controlled by the BNC tube content. Two tubular bioreactors were employed for the processing of BNC tubes.

The tubes produced from the first bioreactor which was assembled with an about 60 mm silicone tube (inner diameter × external diameter: 2 × 3 mm) and a glass tube (8 × 10 mm), were defined as S-BNC tubes. The other bioreactor was consisted of two silicone tubes with various calibers (2 × 3 mm, 8 × 9 mm), were defined as D-BNC tubes. The D-BNC/PVA composite exhibited a higher tensile strength when compared to S-BNC/PVA composite. This criterion is favored for veins and artery transplantation. Figure describes the proliferation of pig iliac endothelium cells (PIECs) on coverslips, BNC tubes, BNC/PVA composite tubes, and PVA tubes at 1, 3, 5, and 7 days after cell seeding.

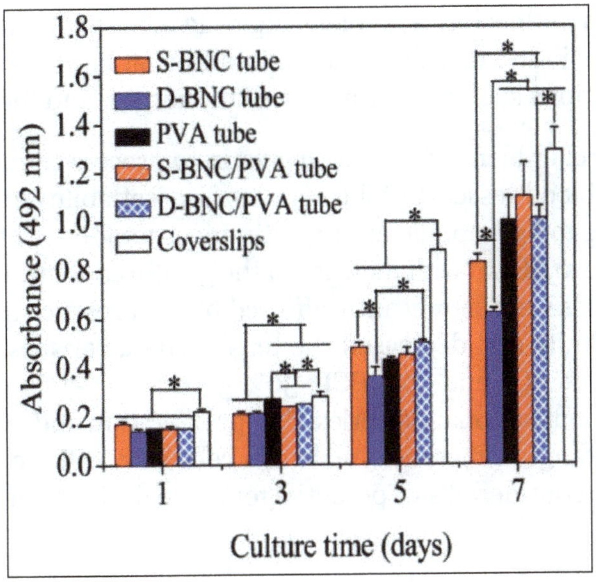

Figure shows proliferation of pig iliac endothelium cells (PIECs) on coverslips, bacterial nanocellulose (BNC) tubes, BNC/poly(vinyl alcohol) (PVA) composite tubes and PVA tubes at 1, 3, 5, and 7 days after cell seeding. Significant variance between groups is shown by asterisks (* $p < 0.05$). The tubes processed from the first bioreactor which was assembled with an about 60 mm silicone tube (inner diameter × external diameter: 2 × 3 mm) and a glass tube (8 × 10 mm), were defined as S-BNC tubes. The other bioreactor was consisted of two silicone tubes with various calibers (2 × 3 mm, 8 × 9 mm), were defined as D-BNC tubes.

Another key point is that the composites of natural biopolymers such as chitosan, gelatin, collagen, alginate, and hyaluronic acid have been applied as implants and wound dressings as well as surgical sutures for decades. Recently, to mimic the tensile and structural characters of biological blood vessels, macroporous blood vessels were developed by researcher through fabricating chitosan/gelatin bilayered composites. On one hand, the macroporous layers displayed a large surface area for cell proliferation and adhesion. On the other hand, the outer nonporous layer offered higher elasticity, flexibility, and cell protection. The biodegradation of the macroporous layer with time was confirmed by fibroblast cell proliferation on the fabricated scaffolds. The viscoelastic nature (elastic (G′) and viscous (G″) moduli) of the fabricated chitosan–gelatin hydrogel was measured at 37 °C. The experiments showed that the scaffolds had pore diameters between 100 and 230 μm, a porosity of 82%, elongation at fracture of 112.5 ± 13%, tensile strength of 95.81 ± 11 kPa, proliferated fibroblasts over 20 days, and 50% in vitro biodegradation after 16 days.

Owing to their low degradability and excellent mechanical characteristics, synthetic polymers such as poly(lactic-co-glycolic) acid (PLGA), PLA, and polycaprolactone (PCL) have been considered as promising scaffolds for constructing cardiac tissue. Researcher represented a platform for creating myocardia and for achieving in vitro cardiomyocyte cultures for drug screening by processing novel PLA/chitosan nanofibrous scaffolds. The scaffolds with random and aligned nanofibers were realized to promote the viability and cardiomyocyte attachment. On the one hand, aligned nanofibers enabled the construction of cardiac tissue through developing cardiomyocyte growth along their longitudinal axis. The processed fibers of PLA/chitosan (7:1) remarkably prepared the extracellular matrix as well as promoting the cell–scaffold bindings. Surface bioactivity plays a vital role in cell proliferation and adhesion. The blending of PCL with gelatin is an interesting concept whereby to overcome their individual limitations, in addition to being a promising approach for tissue engineering applications such as in muscles, skin, nerves, and teeth.

To demonstrate this, researcher blended PCL with gelatin to process composite fibers with improved cell–matrix interactions as blood vessel scaffolds. The surface wettability of PCL fibers was remarkably modified from hydrophobic to hydrophilic after gelatin incorporation. The fiber diameters increased from 1.01 ± 0.51 μm to 1.61 ± 0.46 μm with increase of the gelatin content. The crystallization and thermal resistance of the scaffolds were significantly affected by the presence of gelatin. The high contents of gelatin inhibited crystallinity and decreased the tensile characteristics of the composites. On the other hand, scaffolds of 100% PCL and 70% PCL: 30% gelatin (P7G3-C) were in agreement with the tensile characteristics of human coronary arteries. Cell proliferation and cytoskeleton staining studies revealed the developed bindings of mesenchymal stem cells with PCL/gelatin fibers. To sum up, the P7G3-C composite could be considered as a potential template for blood vessel tissue scaffolds.

Skin

Being the first human body barrier against harmful antigens and pathogens, the skin is considered as one of the most vital body parts and the first immune system organ. When skin is subjected to different stress conditions such as injury and serious burns, skin tissues will acquire secondary infections, necrosis, and damage which cannot be self-healed or repaired. Consequently, developing biocompatible and biodegradable materials for skin regeneration has become a critical issue in the field of tissue engineering. The structural design of skin scaffolds and skin grafts should mimic the structure of the natural ECM.

Molding polymeric materials as 3D nanofibrous structures provides many advantages, such as a high surface area to volume ratio and excellent mechanical properties. Furthermore, polymeric nanofibers can be used as carriers for different drug molecules (e.g., antibiotic, antifungal, and anticancer drugs). Polymer-based hydrogels have been reported as one of the most convenient biomaterials used for this purpose. Several types of hydrogels were fabricated for skin regeneration, being either natural, such as alginate, collagen, hyaluronic acid (HA), and chitosan, or synthetic, such as PVA and polyethylene glycol. Fibrous scaffolds containing collagen combined with drug molecules (Integra, Biobrane, and Terudermis) are commercially available. Recently, researcher electrospun a chemical combination of modified HA and chondroitin sulfate to be applied in skin regeneration and proliferation. In addition, alginate-based thin films were synthesized to be applied in implantation for injured and burned skin.

Oral Tissues

Oral tissue offers a critical function in humans. Dental periodontitis and caries, along with other factors, may lead to damage of the oral tissues. Their wear and tear may result from physical changes while functioning in the environment. The materials inside the oral cavity may be corroded due to exposure to body fluids in the mouth and formation of cytotoxic materials. In vitro tissue engineering of dental tissue has been achieved with significant outcomes.

A tooth structure with enamel and dentin was first regenerated from tooth buds of porcine third molars. These bud cells were added to biodegradable scaffolds and implanted in rats. The tooth structure was attained within 5 to 7 months. Tissue-engineered teeth using porcine tooth buds and a rat model were regenerated by researcher. Polyglycolic acid (PGA) and poly(lactic-co-glycolic) acid (PLGA) were used to cultivate the cells in vitro for six days. These scaffolds maintained the growth of mature tooth tissue. Biodegradable polymer scaffolds were used to produce 3D salivary glandular tissue. In addition, HA and PLA composites were applied to construct temporomandibular joints. These composites were implanted with certain cells to produce osteochondral tissue with vascular bones.

Dental Resin-Based Composites

Resin-based materials have been applied as restorative systems owing to the demand for aesthetic restorations. Resin-based composites are widely applied in dentistry. However, critical alarms still remain regarding their biocompatibility. Therefore, numerous reports have been conducted to measure the toxicology of these materials. These structures are limited by polymerization shrinkage stress and the presence of unreacted monomers. In recent years, novel polymerization processes have been designed to enhance the restoration behavior of these composites.

There are several resin formulation variables such as catalyst type, diluent content, and cure mode. The researchers reported a clear correlation between higher levels of conversion and improved mechanical behavior. The influence of post-curing and long-term aging of these materials on cytotoxicity were reported by researcher. The Ariston pHc (pHc means pH control) was initially moderately toxic, and then became highly toxic for 5 weeks before returning to initial degrees. In addition, estrogen-like materials such as bisphenol A (BPA) and bisphenol A dimethacrylate (BAD) in resin-based dental restorative systems were identified by researcher by using high-pressure liquid chromatography. It was realized that the generic type of materials in resin matrix compositions affected the water sorption and solubility.

Wound Dressing

Many patients are suffering from skin tissue-related devastating trauma. For effective trauma treatment, many studies were achieved to tackle blood loss, physical protection, tissue preservation, and infection. Wound healing is a molecular and cellular behavior that provides necessary characteristics to bypass microbial infections. Growth factor-secreting cells play a vital role to control the wound healing process. In ideal wound healing processes, the fibroblasts transfer to the wound location to fabricate collagen fibers of the extracellular matrix. In addition, the fibroblasts induce the survival and proliferation of keratinocytes. Wound dressings must accelerate the healing process and offer basic protection for the wound location as well. Wound dressings are promising materials that comprise a wide class of materials to provide the latter, including foams, hydrofibers (inter alia), hydrocolloids, hydrogels, films, polymers, cotton, and alginate.

The mechanical and thermal characteristics of different polymers can be improved by incorporating kaolin. Polymeric dressings can be modified and functionalized to display interesting behaviors such as drug release and hemostatic characteristics. Of these, multifunctional hemostatic kaolin–polyurethane (PU) foam composites were synthesized by researcher by incorporating kaolin into mechanically robust PU foam through a one-step preparation. The initial drug release rate and mechanical characteristics were controlled by the kaolin content. At 5 wt % kaolin and less, the composites provided remarkable resilience and good elasticity of more than 140% strain at break. Moreover, foams with 5 and 10 wt % kaolin promoted the hemostatic capabilities. The kaolin content did not influence the absorption behavior of the PU foams. It is noteworthy to mention that these foams can be applied as multifunctional wound dressings as they exhibit different characteristics such as excellent drug release, hemostatic, cytocompatibility, mechanically robust, and absorption properties.

Materials such as chitosan (CS), zeolite, alginate, and collagen were used to achieve localized clotting because of their inherent contact hemostatic behavior. Because of the remarkable antibacterial behavior of titanium oxide (TiO_2) particles, TiO_2 was incorporated as a reinforcing agent to the scaffolds for supporting cell growth. For instance, researcher synthesized a cost-effective and efficient chitosan (CS)/TiO_2 composite membrane for wound healing as well as developing mouse L929 fibroblast cells' survival and proliferation. The membranes were formed with excellent porosity, flexibility, crystallinity, and mechanical strength. The composites offered an excellent antibacterial behavior against *Staphylococcus aureus*. An endpoint MTT assay was applied to test the property of the fabricated membranes to enhance the proliferation and cytocompatibility of fibroblast L929 cells. Additionally, the membranes displayed a greater survival and fast proliferation of L929 cells with reduced apoptosis and oxidative stress.

Banana peel displays promising characteristics such as high mineral concentration, phenolic structure, antioxidant behavior, and antimicrobial performance against yeast, bacteria, and fungi. Recently, researcher has synthesized unique nanocomposite membranes from banana peel powder (BPP) and chitosan. BPP serves as an ionic crosslinker and reinforcing agent for chitosan. The swelling properties of chitosan were reduced after incorporating BPP. Furthermore, glycerol increased the molar volume of the rotating units and relaxation time and consequently reduced the frequency. The effect of frequency reduction was enhanced by increasing the BPP loading. This dip may be attributed to polymer/filler interaction, which increases the relaxation time and the relaxed units. CS–BPP membrane at 10 wt % BPP exhibited the highest activity and sensitivity towards *Candida albicans*. In recent years, nanocellulose (NC) has been used in wound-care dressings because of its significant mechanical strength, water-holding capacity, conformability, and elasticity. NC particles exhibited a high surface to volume ratio, which was utilized to enhance cell proliferation and migration.

However, the NC particles showed poor antimicrobial behavior against pathogenic microorganisms. In detail, a promising bio-nanocomposite from NC, poly(vinyl pyrrolidone) (PVP), and chitosan was fabricated by researcher for in vitro wound dressings. The PVP and chitosan were morphologically miscible and compatible with each other, forming a biocompatible composite with nanocellulose particles via hydrogen bonding between hydroxyl groups of nanocellulose with carbonyl groups of PVP and with amino and hydroxyl groups of chitosan. The results manifested as a significant enhancement in the mechanical and barrier characteristics of PVP/CS/NC films. At 5 wt % NC, the swelling behavior was improved. The tensile strength was increased by 35.6 ± 5.8, 38.4 ± 2.7, and 39.7 ± 6.9 at 3, 5, and 10 wt % NC, respectively. Additionally, blends with 3 and 5 wt % NC offered excellent cytotoxicity, blood compatibility, and remarkable antibacterial performance against gram-negative and gram-positive bacteria.

Another example of natural materials for wound dressing is silk sericin, which is a natural protein derived from the silkworm with significant properties, such as antibacterial, UV, and oxidation resistance, as well as moisture absorption and release. Researcher demonstrated an accelerated wound healing process using silk sericin composites. Researchers discussed the potential of a nanoclay lithium magnesium silicate hydrate (LMSH) cross-linked semi-interpenetrating polymer network (semi-IPN) silk sericin/poly((*N*-isopropylacrylamide)(NIPAm)/LMSH) (HSP) nanocomposite hydrogel as a wound dressing. The hydrogels, which were balanced at 37 °C, exhibited a pyknotic morphology with pore sizes ranged from 0.8 to 1.5 μm, while the cross-section structures displayed a transition from a honeycomb to layered morphology, owing to the improved hydrophilicity.

With increasing sericin content, water vapor transmission rate (WVTR) values increased by 100 g/(m^2·day), while the compression modulus was diminished by 3.859 kPa due to enhancement of the hydrophilic properties. Moreover, the antibacterial behavior was developed after incorporating sericin. Wound healing measurements revealed that the area exposed to a nanocomposite with a mass ratio of 20% (sericin:(sericin+NIPAm)) (HSP20) after 6 days was 83%, three times that of gauze, and was almost recovered after 13 days. This behavior can be attributed to absorption of bacteria by the introduced sericin and adsorption of the decomposed sericin on the surface of bacteria.

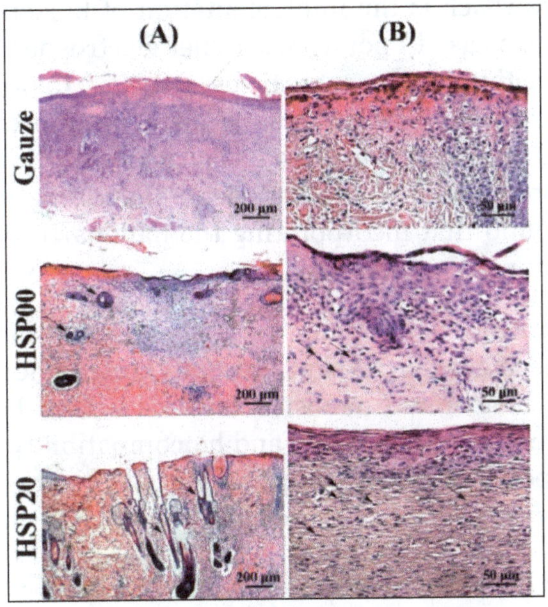

Histological measurement of the skin treated with gauze, HSP00, and HSP20 hydrogel at the 13th day by H&E staining: at (A) 50× and (B) 200×.

The electrospun fibers exhibit superior properties for application as wound dressings, such as the high surface area to volume ratio and porosity, in addition to having excellent biocompatibility with the structure of the extracellular matrix. Additionally, fiber composites show semipermeability, absorption, and hemostatic characteristics. According to this concept, thermosensitive poly(*N*-isopropylacrylamide) (PNIPAAm) was electrospun with poly(l-lactic acid-co-ε-caprolactone) (PLCL) and antibiotic ciprofloxacin (CIF) to process a potent wound dressing. The diameter of the processed cylindrical nanofibers was reduced after introducing CIF. The water angle of the prepared mats was temperature-dependent, and the wettability of the PNIPAAm/PLCL nanofibers decreased at a temperature around 32 °C. In vitro drug release measurements revealed that CIF was released from the mats over 200 h. The drug-loaded mats offered excellent proliferation of L929 fibroblasts and biocompatibility as well as remarkable antibacterial behavior against *S. aureus* and *E. coli*. In vivo wound healing studies confirmed the accelerated healing process effected by the drug-loaded nanofibers.

Additionally, caffeic acid has two vicinal hydroxyl groups on an aromatic ring, defined as a catechol group, and displays different properties, such as antiviral, antioxidation, antibacterial, and anticoagulatory and anti-inflammatory effects. For a wound healing approach, researcher electrospun PCL with chitosan (CS) and chitosan–caffeic acid (CCA) to produce fibers without any bead formation. The average diameters were 1.30 ± 1.07, 1.20 ± 1.22, and 0.94 ± 0.68 μm for PCL, PCL/CS, and PCL/CCA fibers, respectively. In vitro morphology, cell viability, and antimicrobial behavior measurements emphasized that PCL/CCA mats can promote normal human dermal fibroblast-neonatal (NHDF-neo) cell spreading and proliferation with remarkable microbial behavior against *S. aureus* with respect to PCL and PCL/CS fibers. Furthermore, universal tensile machine (UTM) experiments revealed that the tensile characteristics of PCL/CCA fibers were dramatically enlarged compared with PCL and PCL/CS fibers.

Although PCL exhibits unique characteristics such as prolonged decomposition rate and good hydrophobicity and crystallinity, PCL is still not bioactive. One of the most common processes achieved to mitigate this drawback is the implementation of bioactive inorganic additives such as bioactive glasses (BGs). As a case in point, researcher electrospun PCL with BG nanoparticles and silver–cobalt-doped bioactive glass nanoparticles (DB-NPs) to process membranes for wound healing of soft tissue. The weight loss of the fibers was faster after incorporating the nanoparticles. The processed DB-NPs/PCL fibers are considered as promising candidates for soft tissue healing because silver and cobalt ions developed the antibacterial and angiogenic properties of these fibrous membranes. It was found that incorporating the nanoparticles reduced the elongation at fracture and increased the ultimate tensile strength.

Considering the tensile characteristics of skin, these fiber composites may be an alternative for wound healing in such a specific tissue as skin. Recently, researcher prepared SF/graphene oxide (GO) nanofibers with a bioinspired nanostructure for wound healing uses. It was realized that graphene oxide improved the antibacterial and biocompatibility behaviors of SF nanofibers. Table represents different polymer-based composites used in various biomedical applications. From table, it is clearly observed that PCL can be employed in many biomedical approaches.

Table: Summary of different polymer-based composites used in various biomedical applications.

Application		Composite Material
Tissue engineering	Bone	Poly(lactide-co-glycolide) (PLAGA)/calcium phosphate
		Bioglass fibers (BGF) (13–93)/PLA
		PLA/Mg rods
		Polypropylene carbonate (PPC)/poly(d-lactic acid) (PDLA)/tricalcium phosphate (TCP)
		PLA/ethyl cellulose (EC)/hydroxyapatite (HA)
		Alginate (AL)/HA/silk fibroin (SF)
	Blood vessels	PVA/bacterial nanocellulose
		Chitosan (CS)/gelatin
		PLA/CS
		PCL/gelatin
	Skin	Glycoaminoglycoside/collagen
		Modified HA/chondroitin sulfate
		AL/ibuprofen
	Oral tissues	Polyglycolic acid (PGA)/poly(lactic-co-glycolic acid) (PLGA)/porcine tooth buds
		HA/PLA

Wound dressing	Kaolin/polyurethane
	CS/TiO_2
	Banana peel powder/CS
	Nanocellulose/poly(vinyl pyrrolidone) (PVP)/CS
	Nanoclay lithium magnesium silicate hydrate (LMSH) cross-linked semi-interpenetrating polymer network (semi-IPN) silk sericin/poly((N-isopropylacrylamide)(NIPAm)/LMSH)
	Poly(N-isopropylacrylamide)(PNIPAAm)/poly(l-lactic acid-co-ε-caprolactone) (PLCL)/ antibiotic ciprofloxacin (CIF)
	PCL/CS and PCL/CS–caffeic acid (CCA)
	PCL/bioactive glass (BG) nanoparticles and PCL/silver–cobalt-doped BG nanoparticles (DB-NPs)
	SF/graphene oxide (GO)

Sports Applications

With the development of economy, people's living standards improve, more and more modern people relaxing into all kinds of sports venues. And the development of the modern athletic sports in the sports experts focuses on scientific training at the same time, also attaches great importance to the improvement and development of sports equipment. Because of the fiber reinforced composite materials with light weight, high strength, large degrees of freedom of design, easy processing and forming characteristics, obtained widespread application in sports equipment.

Advantages of Fiber Reinforced Composite Material is Applied to Sports Equipment

As is known to all, before the advent of fiber reinforced composites is not, as a sports equipment materials mainly wood, steel, stainless steel, aluminum alloy, etc. Compared with these materials, fiber reinforced composite material has obvious advantages in the following aspects:

- Light: Most of the sports equipment is to rely on human to make the movement, the lighter the better, therefore requires equipment such as tennis racket, golf clubs, bikes, Skis, etc. Fiber reinforced composites in this aspect has the incomparable advantage. Such as the golf clubs made of carbon fiber reinforced materials, adopting the take-up molding in carbon fiber cloth, the mechanical properties of a metal rod improved many, 30% ~ 50% and the weight is lighter than the metal rod.
- Mechanical performance is good: Sports equipment should have good usable performance must have excellent mechanical properties. Fiber reinforced composite material has outstanding than strong. Degrees, modulus and elastic modulus, more suitable for used in sports equipment. Composite material has a good damping shock absorption performance is also one of the reasons its as raw materials for the sports equipment.
- Can design: Composites forming technologies development greatly enhanced its design degrees of freedom than traditional materials, all kinds of products can always find the corresponding. Forming method, thus, according to the different situation of player itself and design respectively, and easy maintenance, low cost.

- Other aspects: In the process of the development of sports equipment, environmental protection and cost performance is also must consider the problem. Composite materials in use process will not. Send out harmful gas, and thermoplastic composite materials can be recycled use, usually at the same time, in the production of composite material prices of raw materials is compared commonly low, low processing cost, conducive to the dissemination and promotion of the product.

Reinforced with Fiber Materials and Fabric Structure

Enhanced with fibre materials mainly include common glass fiber, carbon fiber, synthetic fiber, etc. These fiber materials can be processed into yarn, cloth, belt, used in the form of a mat, chopped strand, etc.

- Inorganic fibers:
  - Glass fiber: Low hygroscopicity, size stability, heat resistance, aging resistance, good chemical resistance and flame retardant, elastic modulus, tensile strength and lower elongation. Glass fiber reinforced composite material is a kind of wide application development earlier, composite material, its tensile, bending, impact strength and stiffness are larger.
  - Carbon fiber: With high strength and modulus, it with the resin poor wettability, adhesion, before the preparation of composite materials, table should be carried out of the fibers. The surface activation treatment. Carbon fiber reinforced composite material is a kind of strength, stiffness and heat resistance are good composite materials.
  - Silicon carbide fibers: Silicon carbide fiber tensile strength, high modulus, good heat resistance, resistance to high temperature of 1250 e, good compatibility. Silicon carbide fiber resin. Matrix composite compressive strength, impact strength and abrasion resistance is better than that of carbon fiber reinforced resin.
- With high strength and high modulus synthetic fiber:
  - Aromatic polyamide fibers: The fiber characteristic is low density, high strength, high modulus, high temperature resistant, and resin has a good bonding effect. It has enhanced the tensile strength of the composite material is better than that of the glass fiber and carbon fiber reinforced material, the modulus of elasticity is 2 times of glass fiber reinforced material, but lower than that of carbon fiber reinforced materials.
  - Ultra high modulus polyethylene fiber: This fiber has high strength, chemical stability and good biological adaptability, low density, but the viscosity difference when used in composite materials. Limit, must pass a plasma ablation, radiation - induced grafting of different methods, such as surface modification processing.
  - Hybrid fiber: Hybrid fiber reinforced refers to two short fiber hybrid or filament one-way enhancement, also can be composed of two different fiber core complex. Yarn as reinforced material, and then with a kind of resin matrix composite. Hybrid fiber reinforced composite materials in addition to the single fiber characteristics, there are some special performance, can meet different application needs.
- Increase with the fabric structure: Fiber reinforced composites reinforced phase in

addition to using the dispersion of short fibers or filament bundle, can also through the fabric forming method. Made of rope, belt, felt and all sorts of different fabric structure, namely the pre in scattered it has no fiber no tangles, shear effect between the fibre and other defects. This estimate can be according to the need of the composite material of the final product, made of different structure and different shapes, such as woven fabric, knitted fabric, woven fabric and nonwoven fabric, multi-layer structure, such as fabric, three-dimensional fabric by these forms of textile structural materials for enhanced phase, don't need to machining, it can be made a molding h-beam, l-shaped beams, cross beams, conical, such as spiral shaped structure.

Made of Fiber Reinforced Composite Material Sports Equipment

There are many different kinds of sports equipment, the following are common fiber reinforced composite material sports equipment to make a simple list, and makes detailed introduction of some products.

Table: Examples of fiber reinforced composite materials application in the sports equipment.

Form	Application
Plate-like structure	Skis, surfboards, windsurfing, table tennis boards, slats and gliding wing spar etc.
Tubular structures	Tennis, badminton, fishing rods, golf clubs, baseball bats, hockey sticks, pole shaft, etc.
Sheet structure	All kinds of helmets, golf club heads, the hull structure of the various boat classes.
Other structures	Match with a variety of vehicles, Sword, climbing ropes, various lines etc.

- The skis: Skiing, snowboarding is related to the athletes' life safety and performance, and the structure of the skis and material is more complex.. General skis have wood, metal and fiber composites (general glass fiber). Wooden light and cheap, but easy to be affected with damp be affected with damp is out of shape. Aluminum alloy metal skis price is higher, the requirement to the snow, high adaptability is poor. Fiber composites of skis is suitable for any types of snow, snow, and easy in maintenance. Skis are typically on the market performance of sandwich composite materials. The skis core material is made from wood or PU, PVC, etc., the elasticity of the skis is from:
    - Carbon fiber is located in the upper core layer, can strengthen the skis flexion degree.
    - Glass fiber in the core layer above, can play a role in connection, can connect the panel and the core layer, increase the toughness of skateboarding, also can let slide more powerful.
- Golf clubs: Shakespear companies in the United States in 1972 by filament winding legal system into a golf club, in the same year, the United States G.B rewer using CFRP (carbon fiber reinforced composite material) made the cue, after that, in order to adapt to the distance and direction of the ball's flight stability requirements, in terms of weight, size and load to improve them. Now high-end golf clubs, using carbon fiber composite material, small density, high strength, high elasticity, impact resistance, make golf clubs can be used repeatedly, but also make the athlete can give full play to the golf ball strength and technology.
- The bicycle: In the mid-1980s, Italy, France, Britain and the United States have developed the stick with carbon fiber tube and aluminum alloy joint. The carbon fiber bicycle into the

frame. Its frame chrome molybdenum steel frame light in weight and strength, stiffness is higher than chrome molybdenum steel frame, once developed, so it is used as a special race car. Well-known German driver, who won men's cycling road race ulrich/mount o is made of carbon fiber reinforced composite material support, the quality is only 715 kg. The current commonly used resin transfer molding (RTM) to batch production bike process.

- Tennis racket: The high, middle-grade tennis rackets in the world today are mostly made of carbon fiber composite materials. The carbon fiber used in tennis racket at the earliest. Carbon fiber composite materials can be big tennis racket, absorb shock absorption performance is good, design freedom. Compared with other materials, carbon fiber used in tennis racket has the following advantages: creates can manufacture large tennis racket: wooden compared with the past, under the same weight, the racket area can increase about 115 times, the tension of cable than the average increased by 20% ~ 45%. DHS absorb shock absorption performance is good: carbon fiber composite vibration damping performance outstanding, it is not easy to start up, also easy to stop after start up.

Civil Applications

While composite materials have been in existence for centuries, the incorporation of composite technology into the civil engineering world is over five decades old. The application of composites in the defense industry dates back to the early 1940's. Several industries started using composites because of their high strength-to-weight ratio and resistance to corrosive elements and weather. The construction industry has been struggling to develop materials that can withstand harsh environments and help build sustainable concrete members. Steel is a widely used concrete reinforcement material which, unfortunately, leads to early deterioration of structures. The introduction of composites into the construction world has given the industry an opportunity to build sustainable concrete infrastructure.

There are not a whole lot of choices when it comes to selecting construction materials. Composite materials such as GFRP fiberglass rebar is the latest technology which adequately addresses many structural issues such as premature deterioration of concrete and structural deficiencies. Traditional materials include steel, wood, and masonry. While some of these materials demonstrate composite properties, they offer limited strength, durability, stiffness, and resistance against corrosive elements. These factors differentiate composite materials from traditional materials.

The scope and applications of advanced composites such as glass fiber reinforced polymer (GFRP) reinforcement have grown significantly over the past few years. These materials have the potential to completely transform the way we construct bridges, rail LRT, mining and tunneling, retaining walls, and other sensitive waterside buildings. GFRP composite materials are designed using two major constituents, reinforcing fibers and polymer matrix. The function of reinforcing fibers is to provide strength and stiffness, while the polymer matrix allows proper load transfer between the fibers and acts as a shield to protect the fibers from environmental elements. Vinylester or polyester resins can be used as a polymer matrix.

So far as the properties of GFRP composites are concerned, few of the most notable are corrosion-resistance, high strength, and lightweight. Composites are also anisotropic which means they have different strength properties in different directions. They can be tailored to meet the complex design requirements for both new and rehabilitation applications. The civil engineering community started using composites in tandem with other traditional materials. For instance, GFRP reinforcing rebar has been used in lieu with traditional steel to strengthen a concrete structure.

Polymer composites are multi-phase materials produced by combining polymer resins such as polyester, vinylester and epoxy, with fillers and reinforcing fibres to produce a bulk material with properties better than those of the individual base materials. Fillers are often used to provide bulk to the material, reduce cost, lower bulk density or to produce aesthetic features. Fibres are used to reinforce the polymer and improve mechanical properties such as stiffness and strength. High strength fibres of glass, aramid and carbon are used as the primary means of carrying load, while the polymer resin protects the fibres and binds them into a cohesive structural unit. These are commonly called fibre composite materials.

Polymer composites have enjoyed widespread use in the construction industry for many years in non-critical applications such as baths and vanities, cladding, decoration and finishing. In 1999, the construction sector was the world's second largest consumer of polymer composites representing 35% of the global market. In recent times fibre composite materials have been increasingly considered for structural load bearing applications by the construction industry and have established themselves as a viable and competitive option for rehabilitation and retrofit of existing civil structures, as a replacement for steel in reinforced concrete and to a lesser extent new civil structures.

Uses of Fibre Composite Materials in Construction

Although the use of structural fibre composites in critical load-bearing applications is relatively rare one of its most common uses in the construction industry is repair of existing structures. The material is also used as a replacement for steel in reinforced and stressed concrete and in very rare cases to produce new civil structures almost entirely out of fibre composites.

1. Rehabilitation and retrofit: Some traditional rehabilitation and retrofit methods use concrete or external steel sheets to re-introduce or improve structure properties such as strength and ductility. The ability of concrete to form complex shapes and its suitability to submerged installation has seen it used for encapsulation of elements such as bridge piers. Steel can be bonded or bolted to deteriorated concrete structures to provide strength and stiffness improvements with relatively little additional weight. In the last decade the number of instances of fibre composites used as a surface layer that either protects and/or improves on the response of the encapsulated element has been increasing.

In these cases the materials are usually bonded externally to the structure in the form of tows (fibre bundles), fabrics, plates, strips and jackets. The advantages offered by composites in these forms include their ability to bond well to many substrate materials and to follow complex shapes. Composites also offer a potential benefit over isotropic retrofit materials, such as steel, by allowing enhancement of strength without increasing stiffness and vice versa.

2. Concrete structures reinforced with fibre composites: Concrete reinforced with fibre reinforced polymer (FRP) materials has been under investigation since the 1960's. Unstressed FRP reinforcement has been developed in a number of forms including ribbed FRP rod similar in appearance to

deformed steel reinforcing bar, undeformed E-glass and carbon fibre bar bound with polyester, vinylester or epoxy resin, E-glass mesh made from flat FRP bars and prefabricated reinforcing cages using flat bars and box sections. Stressed FRP reinforcement is also available, usually consisting of bundles of rods or strands of fibre-reinforced polymer running parallel to the axis of the tendon. These are used in a similar fashion to conventional steel tendons.

The durability performance of FRP reinforcements is considered by some to offer a possible solution to the problem of corrosion of steel reinforcement, a primary factor in reduced durability of concrete structures. Other reported advantages of FRP rebar include enhanced erection and handling speeds and suitability to applications which are sensitive to materials which impede radiowave propagation and disturb electromagnetic fields.

3. New fibre composite civil structures: A small number of new load bearing civil engineering structures have been made predominantly from FRP materials over the last three decades. These include compound curved roofs pedestrian and vehicle bridges and bridge decks, energy absorbing roadside guardrails, building systems, modular rooftop cooling towers, access platforms for industrial, chemical and offshore, electricity transmission towers, power poles, power pole cross-arms and light poles and marine structures such as seawalls and fenders.

The potential benefits offered by fibre composites include high specific strength and specific stiffness, tailorable durability, good fatigue performance and the potential to reduce long-term costs. However, in many cases these potential benefits are difficult to realise and are sometimes based on specious fact and irrelevant data. In addition to this, the lack of bona-fide applications has caused many constructors to be sceptical of the material's ability to provide a viable alternative to traditional materials. Many of the existing applications are experimental in nature and are aimed at demonstrating the ability of fibre composite materials to perform in certain applications. To this end they may be successful in terms of structural performance, but offer little by way of meaningful financial performance data.

Issues Slowing the Adoption of Fibre Composite Materials in Construction

There is little doubt that fibre composite materials are structurally capable, however their ingress into the construction industry has been slow to date. Issues slowing the adoption of fibre composite materials in construction are issues of cost, structural performance and durability.

Cost

Cost can be considered in terms of short-term costs, such as design, construction and installation, and long-term costs such as maintenance, modification, deconstruction and disposal. These can be further grouped into direct costs, such as materials and production, and indirect costs, such as interruptions to traffic, depreciation, resale value and impact on the environment. Difficulties can arise in determination of the lowest cost of competing solutions when interested parties place different levels of importance on these different types of cost.

1. Short term costs of fibre composites: Currently, fibre composite materials are expensive when compared to conventional construction materials on an initial cost basis. There are a number of factors contributing to the high cost of composite materials including; high cost of raw materials

and processing, the use of imported materials, the general acceptance of high prices in markets such as marine and aerospace and occasional low availability of material.

In line with the evolution of other composites uses, such as sporting equipment, some researchers believe it is likely that production volume increases resulting from the use of fibre composites in civil engineering applications will lead to decreased cost of materials. However, locally this may be an optimistic outlook as the majority of fibre composites materials used in Australia are imported and are therefore subject to a range of international economic variables such as fluctuations in overseas production costs, transport and import costs and fluctuations in the exchange rate between Australia and countries such as Europe, United States and Japan, which supply us with carbon, aramid, E-glass fibres and many resins. When this is considered in conjunction with the tendency of suppliers to provide price reductions based on quantity purchased, accurate costing of an FRP construction project can seem difficult and suggest that it could be some time before anticipated price drops could significantly influence project cost.

2. Fabrication cost: In addition to relatively high material costs, the short-term cost of FRP materials is dependant on fabrication. Most fibre composite manufacturing techniques were originally developed for the aircraft, marine and/or car industry. The construction industry is vastly different to these, as constructors tend to be concerned with the design and construction of rather large-scale structures. Also, design specifications usually differ from project to project and therefore very little duplication of design solutions occurs. As a result most construction projects tend to be 'one-off' jobs. This situation is in contrast with the manufacturing industries, where mass production of one design solution is common. As a result, design and manufacturing methods that are highly successful in the manufacturing industry are often not viable in civil engineering.

Some short-term costs, such as transport and erection may benefit from production of large, lightweight, modular components. Lower weight can translate into reduced transport and cranage costs, while the use of fewer large modular components can reduce erection time. Researcher points out that although it is difficult to quantify indirect savings, they have a cost that is present. He believes that savings can be accrued at the systems level due to faster construction thereby causing less distress and disruption to the community, lower dead weight requiring smaller and lighter substructure as well as lighter construction equipment. Others claim that road closure in busy areas, traffic congestion, detours, environmental costs, administrative costs, downtime in industrial applications, and reduced maintenance can represent substantial benefits.

3. Costing of fibre composite materials: Techniques used to cost civil infrastructure projects can vary from project to project depending on individual circumstances. Often times, parties with financial interests in building projects will base cost decisions on the initial cost of the structure. This is primarily due to project budget limits and the tendency for the owner to be more concerned with obtaining the best structure possible for their money and less concerned with its long-term performance.

Two techniques claim to consider most of the critical issues in the application of composites in civil applications. The first has been developed by El-Mikawi and Mosallam and uses the Analytical Hierarchy Process (AHP) to provide various levels of focus to evaluate options for project needs, management, manufacture and maintenance. However, their method does not allow comparison of tangible and nontangible factors, such as impact on amenity, or cost/benefit analysis of the options important in a realistic evaluation of composite materials.

The second approach is known as the Whole of Life (WOL) approach, which is derived from lifecycle costing. WOL costing appears to be more comprehensive and includes initial cost, maintenance cost, operating cost, replacement and refurbishment costs, retirement and disposal costs and other costs such as taxes, depreciation and additional management costs. Ongoing costs are allocated a value based on the expected lifespan of the building. This method could be used to compare, in economic terms, the advantages of new and existing materials in structures which are designed to equal performance criteria such as minimum service life, strength and stiffness.

Structural Performance

1. Specific strength and specific stiffness: Fibre composite materials are often claimed to offer potential benefits to construction projects through high specific stiffness and high specific strength. This is primarily based on the ability of materials that possess high specific strength and high specific stiffness to produce structures with low self-weight in some applications. The potential benefits arising from low structure weight include the freedom to produce larger structures, larger components in the factory, reduced transport and erection costs and reduced size of substructure and foundations.

Judicious use of some fibre composite materials instead of traditional construction materials could potentially produce a lighter structure and lead to a number of cost savings, but these may be difficult to realise. Currently a proprietary FRP bridge deck can cost as much as ten times that of traditional precast concrete planks for a single span two-lane road bridge. The use of a lighter deck could result in transport costs decreasing by as much as 75%. However, the saving would probably amount to a few thousand dollars for installations close to the point of manufacture. Small cost savings may also be made through reduction in lifting costs and savings on concrete substructure, although these would not be considered significant in most cases.

2. Low stiffness: Civil structures are commonly governed by stiffness performance and the use of materials with low gross elastic moduli, such as FRP materials, can cause these structures to become significantly over-designed for strength, making them economically uncompetitive. Whilst reasonably stiff laminates can be produced using carbon fibres, they tend to be expensive and have a lower failure strain which limits their usefulness. As reinforcing fibres evolve it is possible that higher modulus laminates will be produced at a lower cost and which won't require expensive consolidation techniques to increase fibre volume fraction to give higher gross moduli.

3. Tailorable mechanical properties: It is often claimed that fibre composite materials offer designers of civil structures increased versatility over traditional materials through freedom to engineer the material as part of the design process. This may be achieved by varying the type of fibre and resin and the orientation and location of reinforcing fibres to produce structures with a combination of performance characteristics such as strength, stiffness, durability, impact and fatigue well suited to a particular application, potentially using material more efficiently.

But this level of tailorability in fibre composite materials requires a fabrication process that allows localised variations of laminate composition. This tends to preclude the use of some automated procedures, such as pultrusion, which allows some variation in reinforcing fibre type on a ply-by-ply basis, but it does not allow variation in resin composition or localised changes in laminate layup.

Durability

Enviro-mechanical durability is often cited as a key advantage of FRP composites materials over traditional materials. The large range of constituent materials potentially allows design of a material which exhibits very good resistance to long-term deficiencies which can result from environmental influence of moisture, ultraviolet radiation, chemical attack, dynamic loading, freeze thaw cycles and deterioration of material properties through physical aging.

However, fibre composites have been used in civil engineering applications for a relatively short period of time and full understanding of their durability is yet to be achieved. Liao showed that substantial consideration has been afforded to the durability of FRP composite materials in infrastructure applications. However Karbhari identified a lack of long-term data relevant to civil structures with a service life of 75 – 100 years. An international study undertaken by the Civil Engineering Research Fund (CERF) to bridge gaps in long-term durability data related to civil engineering applications identified areas which are lacking such as; moisture effects, alkaline solution, fatigue, creep and physical degradation.

A Long-Term View

Although an enormous effort is underway to migrate fibre composite materials to construction applications they appear to be struggling to compete with traditional construction materials. However, some fibre composite materials may be preferred over traditional construction materials as environmental sustainability becomes more important in the long-term. The "long-term" could be more than fifty years if steel and reinforced concrete are any indication of the construction industry's attitude towards new materials. This would provide a significant amount of time for issues of environmental sustainability in construction to mature and is likely to see a change in the criteria with which evaluation of material options is undertaken. It is likely that material choice will become more focussed on environmental issues such as reduced use of non-renewable natural resources and lower embodied energy.

This trend towards increased environmental responsibility for materials is already occurring in other industries. For example, the European automobile industry is taking steps towards adoption of a "cradle to the grave" (or the "cradle to the cradle" if materials are recycled) philosophy for material use. It is not unreasonable to suggest that a similar model could be developed for the construction industry. If such as model were to be adopted then much more emphasis would need to be placed on the use of materials which do not use non-renewable natural resources and which are recyclable or biodegradable.

In terms of construction material choice, the "green" requirements would be in addition to practical material characteristics such as stiffness, strength, affordability, durability, versatility and easy of use. Our three traditional construction materials each possess some of these characteristics, but none possess all. The ultimate goal would be to develop a material that not only possesses the basic requirements of construction materials, but also the characteristics of an environmentally sustainable material. Composite materials may offer a solution. Their unique tailorability allows virtually any combination of properties, but the cost of this versatility is high.

However, of significantly greater concern is that the current range of materials cannot be considered environmentally sustainable, despite some claims by the composite industry to the contrary.

For example, the production of reinforcing fibres such as glass, carbon and aramid, requires an enormous amount of energy and, in the case of glass is derived from a non-renewable resource. Resins are sometimes claimed to be more environmentally friendly than materials such as metal because they are derived from by-products of the petroleum industry. However, the petroleum industry itself is not sustainable. Similarly some filler materials used to provide bulk to some resin systems are derived from the waste of coal-fired power stations; this industry may not be sustainable in the long term due to public awareness of pollution despite vast coal reserves.

A feature of composite materials is their ability to combine two or more basic constituents to produce a bulk material with characteristics improved over the base materials. This may allow a suitable construction material to be produced from basic materials that are environmentally sustainable. Bio-composites, or bio-fibres, is not a new concept; natural fibres such as linen, cotton, hemp and straw have been used as reinforcing materials for many years. On the other hand, natural matrix materials with suitable properties may present more of a challenge. A small number of natural resin products are available, such as rubber, plant resin, and animal glues. However these generally lack the processing and performance characteristics sought after in a matrix resin. This is not to say that the effort spent on development of current fibre composite technology is being wasted. On the contrary, it will be critical to facilitate the migration to "bio-composites" technology in the future. Methods, systems and standards need to be developed which are generally applicable to fibre composite materials in the construction industry and which will be applicable to new types of fibre composite materials as they become available.

By the time that natural materials and associated design methods are sufficiently mature to allow their widespread use, issues related to construction material sustainability are likely to have become paramount in material choice. The coincidence of these three factors could see natural composite materials at a distinct advantage over traditional materials. There is a significant research effort underway to develop natural materials and explore their use in construction applications. This research needs to continue in conjunction with development of conventional composite materials in order to provide a solution in the future which will allow wider use of the natural composite materials by the construction industry.

Offshore Applications

In surface transportation, reinforced plastics are the kind of composites used because of their huge size. They provide ample scope and receptiveness to design changes, materials and processes. The strength-weight ratio is higher than other materials. Their stiffness and cost effectiveness offered, apart from easy availability of raw materials, make them the obvious choice for applications in surface transportation. In heavy transport vehicles, the composites are used in processing of component parts with cost effectiveness. Good reproductivity and resilience handling by semi-skilled workers are the basic requirements of a good composite material. While the costs of achieving advanced composites may not justify the savings obtained in terms of weight vis-a-vis vehicle production, carbon fibers reinforced epoxies have been used in racing cars and recently for the safety of cars.

Polyester resin with suitable fillers and reinforcements were the first applications of composites in road transportation. The choice was dictated by properties like low cost, ease in designing and production of functional parts etc. Using a variety of reinforcements, polyester has continued to be used in improving the system and other applications. Most of the thermoplastics are combined with reinforcing fibers in various proportions. Several methods are used to produce vehicle parts from thermo plastics. Selection of the material is made from the final nature of the component, the volume required, apart from cost-effectiveness and mechanical strength. Components that need conventional paint finishing are generally made with thermosetting resins, while thermoplastics are used to build parts that are moulded and can be pigmented. Press moulded reinforced polyester possess the capability to produce large parts in considerable volume with cost-effectiveness.

In manufacturing of automobile parts, glass and sisal fibers usually find the maximum use. Sisal costs very less and this alone has prompted extensive research to come up with applications in which sisal is the dominant reinforcing material in filled polyester resin, in parts where specific mechanical properties are required and appearance is not very important. Heater housings, which find uses for sisal, are produced by compression moulding. Since a variety of glass fibers are available, it is used as reinforcement for a large range of parts of different types. Rovings, non-woven mats are the commonly used low cost versions. Woven cloth is applied in special cases, where particular properties are required as cloth is not known to be amenable to large quantity production methods. Since the automobile industry is replete with models, options and changes in trends, the material selection and combinations offered by the materials is also wide-ranging. Along with a measure of conservation, the choice is also dictated by the demands of the competitive market for new and alternate materials.

A reinforced-plastic composite is likely to cost more than sheet steel, when considered on the basis of cost and performance. In such a case, other qualities must necessarily justify the high expenditure. Mechanical properties of the parts, which affect the thickness and weight, must offer enough savings to render them more effective than steel. It however shows a higher machining waste than reinforced plastics. The fabrication costs of reinforced plastics is controlled by the devices and tooling used for producing them. In turn, it is dependent on the basis of the quantity of components needed. Some complicated parts of light commercial vehicles, which need casting, may be compression moulded from composites of the sheet or bulk variety. State-of-art technologies of moulding, tooling and fabricating have thrown open possibilities of increased manufacturing of vehicles that use reinforced polyesters.

Materials used in automotive body parts show high tensile strength and flexural moduli. The material is not ductile and hence will not yield and the failure is accounted only in terms of fracture. These properties and thickness, determine the maximum bending moment which is several times higher than the point of fracture for steel sheets. Reinforced plastics can be given the metal finish, although the cost of achieving this continues to be prohibitive. They are restricted in their use in car components. While the defects in painted sheet metal parts are easily overlooked, the fiber pattern texture is obvious, though the surfaceroughness measurements report that it is smoother.

In commercial vehicles, appearance is also important as is the functional aspect. Since a commercial vehicle is more a capital investment, it is the returns from such investment that are considered. The rate of return depends on initial cost, durability and maintenance costs. Reinforced plastic is a boon in the sense that it uses shorter lead times and tooling cost is considerably cheaper. Commitments to launch a new model are kept easily, since the time between production and introduction can be co-ordinate perfectly. Studies have shown that composite panels may be used as the complete outer skin of the body to give a unique look. Sheet moulding compounds of resins are most suited for this purpose. Inner and outer reinforcing is done by panel assembled by adhesive bonding and riveting.

Good stability against corrosion or impact makes the composites widely used in vulnerable valance panels below the front and rear bumpers. Signal lamps, indicator lamps of vehicles are fabricated from glass-reinforced composites and tractors have a different selection methodology from that of passenger cars. The most crucial parameter is weight reduction as it directly affects efficiency, payload and the economy. Durability is the chief factor as these vehicles are normally realizations of capital investments. Time required, cost and frequency of maintenance add substantially to the total costs. Therefore it is natural to try and reduce these factors to a minimum. Fiber glass reinforced polyester is widely used in various parts of trucks. The fatigue properties of the materials and the low weight, ability to sustain strains from the engine heat and low frequency road vibrations are features that favour composites in trucks and other heavy vehicles. Reinforced plastics do not cost too much to tool, and they are now extensively used for automobile parts, indicator and signal lamp parts and other accessories. Truck bodies and trailers use assemblies and parts made from reinforced plastics to a great extent. The use of light metals, which lends itself to simple shapes and extrudable forms, is also found to be economical. The low heat transfer coefficient of composites enables their use in refrigerated units. Glass reinforced polyester has all the properties that make it ideal for this purpose and has become the standard material.

Plywood panels laminated with thin layer of reinforced plastic are also widely used in truck carriages. Several methods are used to produce them and the low cost, and strength offered by plywood make it appealing. They may not be as light weight as desired, but are found to be considerably more durable than the conventional constructions, particularly in intensive service conditions. There is also a growing market for products like pick-up trucks and trailers which are made using glass reinforced plastics at competitive prices. Vehicles meant for consumer markets tend to be designed attractively, their appearance is different and the tooling is easy. Reinforced plastic moulds from the full size model of the part are taken. The pattern may be either made from wood, clay or plaster. Sheet metal body of automobile is moulded in matched moulding dies with fiber glass or rovings as reinforcements. Large, complex parts can be produced and their reproduction can be made equally fast too. The parts made using this technique are generally of a uniform thickness, because the extra reinforcing fibers cannot be easily or accurately loaded.

Moulding compounds are used when small and rigid parts have to be made and they are usually prepared by casting. The bulk moulding compound is composed of polyester resin, glass fibers and filler. Strength is lesser than that obtained by other moulding processes. Moulding compounds can be prepared as continuous sheets. Due to the longer fibers and high glass content, their strength is better than those made in bulk moulding method. Injection moulding and pultrusion are used to a limited extent to produce materials that may be applied in structural materials. After separate fabrication of parts by moulding or other methods, assembling the vehicles body has to be done. Tooling and fixtures similar to those used in metal parts are used for piercing and drilling operations.

Assembling of various parts is usually done by adhesive bonding, using resins that are catalyzed to cure at room temperature in a short time. Good bond strength is achieved even without pressure, although clamping fixtures are needed to locate the assembled parts correctly in relation to each other. High build primers are used to give high gloss finish so that surface imperfections do not become obvious. Commercial aircraft applications are the most important uses of composites.

Aircraft, unlike other vehicles, need to lay greater stress on safety and weight. They are achieved by using materials with high specific properties. A modern civil aircraft must be so designed as to meet the numerous criteria of power and safety. Glass reinforced composites are the most desired materials as a result of advanced technology that has gone beyond the design and application. In cases where high moduli of elasticity values are less important, fiberglass is the natural option because of the low cost of material. The matrix material used with fiber glass, however, limits its use to low temperatures, usually below 121 °C, although it is not a debilitating limitation for the fiber, as its properties can still be used and maintained at temperatures beyond 426 to 482 °C.

Fiber epoxy composites have been used in aircraft engine to enhance the performance of the system. The pilot's cabin door of aircrafts has also been made with fiber glass resin composites and these are now used in other transport systems. The boron-graphite materials were initially designed for fighter aircraft components and their use in commercial aircraft has been very less. There are a few instances of applications of these composites in wide use currently although experimental applications are several. They are presently limited to secondary structures which can be used in commercial aircraft with considerable safety. The data from such experimentation on the long term effects of loads and stress on the structure provides input for design. Both dynamic and static conditions are combined in the turbojet engine and research has always been directed towards this. These applications involve lightweight materials and this combination offers advantages. The weight of the rotors, compressors and bearings are reduced. Initially, turbojet engines were used in fighter aircraft and later in commercial planes. The need of a commercial plane is long service life and durability.

Some of the turbofan engines are designed to meet the manifold requirements of transport sector. The engine can be improved by improving the efficiency of propulsion or reducing the weight. The notable stiffness and strengths of composites permit reduction in the number of compressor stages by higher blade loading. The use of composites in rotors, compressors and engine parts are estimated to lead to weight savings.

Aeronautical engineering comprises of various distinct areas that produces vehicles capable of performing distinct flight programmes. Initially importance was given to weight, speed and power, but other parameters that influence market acceptance of the aircraft should also be considered during design. These conditions call for selection of materials that give less than optimum efficiency in terms of structure and systems. Hence, it is important to consider performance needs as well as service properties. Airframe design starts with evaluation of flight conditions which the aircraft will encounter. In recent designs, wind tunnel tests and analysis are being done to determine the lift and drag forces. Once determined, they are used to develop various related factors of structural engineering. The selection of material, it follows, enters naturally into the picture at the early stage of design itself. The high strength of composites allows designing of higher aspect ratio wings in aerofoil sections.

Nowadays, composites are used in peripheral structures of aerodromes. Conventional constructions of composites ought to cost much less in future and will not be a constraint. Automation along with high standard for filament and matrix materials will also decrease fabrication costs, as the rejection on grounds of quality will be less. Performance, reliability and efficiency of operators alone can assure the success of any programme and the space program in particular. The potential and application of high performance composites has revolutionized space structural technology.

Glass filaments have been used in space vehicles for several years now as laminations in secondary structures. As far as its importance as an engineering material is concerned, glass is attractive because of high specific strength, low cost, good forming characteristics, high impact resistance and thermal stability. Thanks to its insulating property, glass has been used to a great extent in thermal isolation components and structures. However, a drawback with glass is its low elastic modulus. Although glass is easily handled, it results in excessive tool wear while machining operations are conducted.

Graphite is a widely available economical reinforcement material with high stiffness, high modulus, high strength and high theoretical efficiency. However it is difficult to transfer load between layers. In addition, its compressive properties are less than its tensile properties. The positive points are that graphite has extremely good machining and forming capabilities and has a very low coefficient of expansion. It can also be easily woven into cloth and have high dimensional stability.

Generally, flaws in composites affect the degradation of the static strength more than in metals, but are unlikely to be as detrimental in fatigue loading. The strength is dependent on the shape of the flaws and the critical failure stress in the composite. By using glass fibers as doublers, stress softening can be done around cut-outs. Joints often decide performance and reliability of the composites. Since composites do not exhibit good bearing strengths, conventional joints like mechanical fasteners cannot be used.

Moreover, as reinforcement do not fuse at high temperatures, welding is not very effective. There are three methods of joining composites. In scarf joint, the laminate layers are laid up against wedge shaped metal edge members. Bolt holes are placed through the metal and loads are transported in shear to composites held by adhesive along with them. Metal double-step lap joint has some layers in 90 degree orientation butted to the joint ends. Metal shims are interlaced and inserted and then bonded between composite layers. The load is borne by the shims which subsequently shifts it to the composites. But this has results in thick joints and displaced laminates.

A design concept developed to reinforce certain chosen areas of metal structures for localized strength has several advantages including significant reliability, cheaper in cost, significant weight reduction etc. This means that the risks that are associated with composite problems are minimal. This method also permits the use of conventional metal joints which include riveting and welding. Hence the joints are fairly reliable. A major concern in management of space vehicle systems is the delivery of hardware that meets the important requirements of schedule, performance and cost. Increased reliability and performance can be achieved by changing the material among other options. Composites have great potential in this respect and performance can be achieved by extracting weight reduction of structure.

An important consideration in the use of composites is lightweight. Research studies of specific components have shown that using all composite structures a saving of 20 to 45% can be achieved while selectively reinforced metal structures offer about 10 to 25% only. Sometimes weight reduction is required to maintain the center of gravity of the system. Specially, tail weight can be reduced by application of composites so that weights do not have to be added at the nose to maintain the center of gravity position. Composites are selected to solve problems of critical payloads or balance. Few cost factors considered are the initial cost and life cycle cost. In the first case, the cost of generating data, developing new routes and inspection techniques etc., have to be considered. Composite designs have been developed showing ample success in lowering manufacturing costs which recommend them over metal designs. It applies especially to composites with filament reinforcements, which use simpler tools and reduce time taken for fabrication process.

In railway carriages it is desirable to reduce the weight of rail car bodies as well as heavy transport vehicles, which in turn reduce power and braking requirements. It also reduces maintenance costs. Vehicle type, colour selection is the major aspects of building rail cars. Structural concepts on certain aluminium and steel vehicles which are designed from sheets and stiffened by extrusion are not always the most efficient in case of collision. Fire retarding properties, noise and heat insulation and crack resistance are additional qualities. The high cost of material has restricted the use of fiber reinforced plastics in freight cars. Composite materials, however, have been used for component parts such as fasteners, levers, hatch covers in hopper cars etc. Glass fiber reinforced plastic, used for containers, is also used for paneling in rail cars. Fiberglass has been the flexible insulation material of choice for these vehicles.

The principal design criterion for transport vehicles involves conditions that the structural weight must be minimized to conserve raw material, tribological factors and at the same time remain the least polluting. The lifetime should be long and reliable. Reliability should be considerably improved to conserve materials and reduce maintenance costs. Designs should be considered giving importance to fatigue and fracture resistance. They should improve the level of safety and performance. The impacting absorption should be better than those of other vehicles and should have insulation against fire and smoke retarding capabilities.

Glass reinforced systems are used in many city transit cars. The accessories are made from cellular cellulose acetate foam cores, over laid with fiber glass for strength and rigidity. Few vehicles use glass-reinforced polyester for panel dividers. Structurally the exterior of these vehicles combine fiber glass moulded parts over truss-type structures. Every structural design has to inevitably pass the economic viability criteria. These vehicles usually involve preventive maintenance and overhead expenses. While assessing cost impact of the design in question, the total life cycle costs and the computations must relate the improvement to the operating and ticket costs per passenger-mile.

Engineering Applications of FRP Composites

Application of Composites in Aircraft Industry

The use of fibre reinforced composites has become increasingly attractive alternative to the conventional metals for many aircraft components mainly due to their increased strength, durability, corrosion resistance, resistance to fatigue and damage tolerance characteristics. Composites also provide greater flexibility because the material can be tailored to meet the design requirements and they also offer significant weight advantages. Carefully designed individual composite parts, at present, are about 20-30% lighter than their conventional metal counterparts. Although all-composite airplanes are now available in the world market, yet advances in the practical use of composite materials should enable further reduction in the structural weight of airplane. The composite materials used in aircraft industry are generally reinforced fibres or filaments embedded in a resin matrix. The most common fibres are carbon, aramid, glass and their hybrid. The resin matrix is generally an epoxy based system requiring curing temperatures between 120° and 180 °C (250° and 350 °F).

The first structural composite aircraft components, which were introduced during 1950-60, were made from glass fibre reinforced plastics. These components included the fin and the rudder of Grumman E-2A, helicopter canopies, frames, radomes, fairings, rotor blades, etc. Due to high strength and stiffness combined with low density, composites like Boron Fibre Reinforced Plastics (BFRP) and Carbon Fibre Reinforced Plastics (CFRP) .were preferred instead of aluminium for high performance aircraft structures. For lightly loaded structures, Aramid Fibre Reinforced Plastics (AFRP) which possess low density, have been used. The use of AFRP continues to be restricted to the lightly loaded structures due to the fact that although these fibres possess high tensile strength, they have very low compressive strength.

For light aircraft and lightly loaded structural components, Glass Fibre Reinforced Plastics (GFRP) has become one of the standard materials. Over the years, use of composite materials has also increased from few small access panels and canopy frames to almost complete airframe surfaces thereby providing weight savings leading to improved performance, reduced drag and also improved durability and corrosion resistance. Consequently, now-a-days, composite materials like GFRP, CFRP and AFRP have become standard materials for flight control surfaces, engine cowlings, fairings, radomes, landing gear doors, floor panels, fan ducts, etc. in aircraft application. Weight savings provided by composites vary considerably with the type of aircraft and component. Weight savings, in terms of composite weight fraction are shown in figure. These tend to decrease as the overall composite weight fraction increases.

Composite aircraft components used for structural application are generally fabricated with a sandwich construction having face sheets of carbon fibre or carbon fibre combined with aramid or glass fibres with a honeycomb core. For interior aircraft applications, composite parts are required to meet mechanical properties and processability requirements. In addition, materials used within pressurized portion of aircraft must meet the flammability resistance requirements. Interior parts such as overhead bins, sidewall panels, ceilings, floor boards, galleys, partitions, cargo floor board liners, etc. are made of composite materials which are generally fibre reinforced epoxy or phenolic resin.

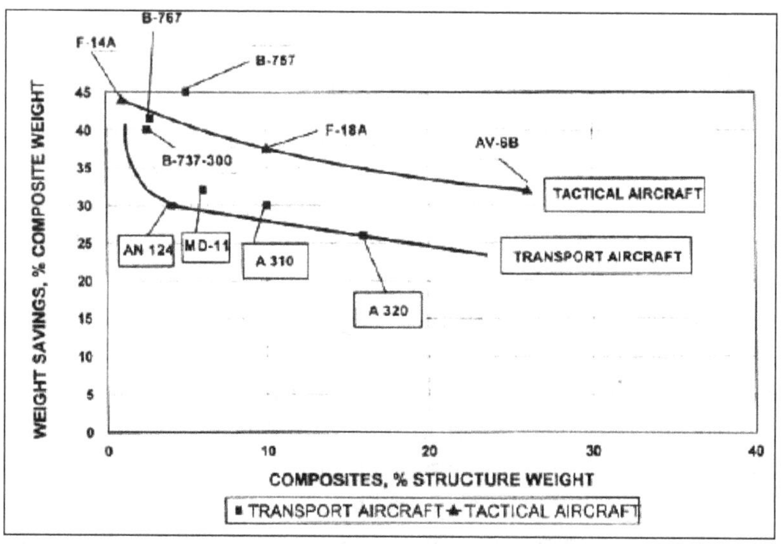

The comparison plot to find weight saving with respect to type of Aircraft.

For interior applications, phenolic resin system is mainly used because of its excellent fire resistant properties including low flammability and low smoke and toxic gas emission. The predominant design considerations for interior components are impact resistance, stiffness and surface smoothness. The commonly used fibres for interior aircraft application are glass fibres. In the present day scenario, use of advanced composite materials has been extended to a large number of aircraft components, both structural and non-structural, based on various factors namely, in-service loading, environmental conditions, etc.

Typical Applications in Aircraft Industry

Civil Aircraft Applications

Airbus Industries used advanced composites on the Airbus A300 aircraft which first flew in 1972. The composite material was used in fin leading edge and other glass fibre fairing panels.

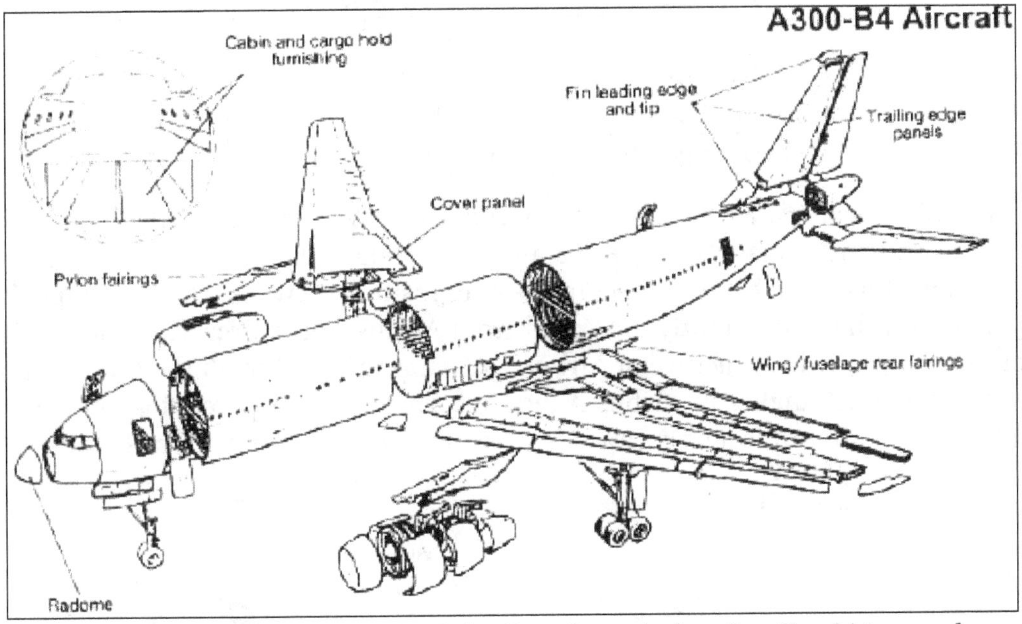

The composite material was used in fin leading edge and other glass fibre fairing panels.

In 1979, the in-service evaluation of Airbus A300 aircraft led to further use of composite components on Airbus A300 aircraft namely CFRP spoilers and rudders, air brakes, CFRP landing gear doors, etc. Use of composites was extended to Airbus A310 aircraft during 1980-85, and thereafter to Airbus A320 aircraft in 1987. Various composite components used on Airbus series of aircraft are given in the table.

Table: The components used on Airbus series of aircraft.

Aircraft Type	Components Made of Composite Materials
Airbus A300B2/B4	Radome, fin leading edge and tip, fin trailing edge panels, cabin and cargo hold furnishings. Fairing -pylon, wing/ fuselage rear.
Airbus A310-300	Rudder, elevator, vertical stabilizer, spoilers, cowl (inlet & fan), thrust reverser, main & nose landing gear door of wing leading & trailing edge panels, nacelles. Fairings -Ion, flap track, win fuselage.

Airbus A320/ A319 & A321	Aileron, horizontal and vertical stabilizer, elevator, rudder, spoilers, flaps, engine cowl, radome, landing gear doors (main & nose), floor panels, wing panels (leading & trailing edge), other access panels, nacelles. Fairings -flap track, wing/fuselage (forward & rear), and main landing gear leg.
Airbus A330	Ailerons, rudder, flaps, spoilers, elevator, horizontal and vertical stabilizer, wing panels (leading & trailing edge), landing gear doors (main & nose), nacelles. Fairings -flap track, wing/fuselage (forward & rear).
Airbus 340	Ailerons, rudder, flaps, spoilers, elevator, horizontal and vertical stabilizer, wing panels (leading & trailing edge), landing gear doors (main & nose), nacelles. Fairings -flap track, wing/fuselage (forward & rear).

Table: Weight of composites used on different aircraft of different Airbus.

Aircraft Type	Weight of Composites
Airbus A300	≈ 4000 lb. (5%)
Airbus A310	≈ 7400 lb (7%)
Airbus A320	≈ 9000 lb (15%)
Airbus A330/A340	≈ 16000 lb (12%)

On Airbus A300B2/B4 aircraft, only glass fibre structures have been used. AFRP and CFRP structures have been used on Airbus A310-200 aircraft. For Airbus A320 and Airbus A330/340 aircraft, GFRP, CFRP and AFRP have been utilized in the composite structures. Use of composites in Airbus family of aircraft has consistently increased from Airbus A300 aircraft onwards. Composites account for about 15% of the structure of Airbus A320 aircraft. Although the total weight of composite structures is much higher in Airbus A330/A340 aircraft, the percentage of weight in relation to total weight of aircraft is nearly 12%. Use of composites on A340 aircraft is shown in figure.

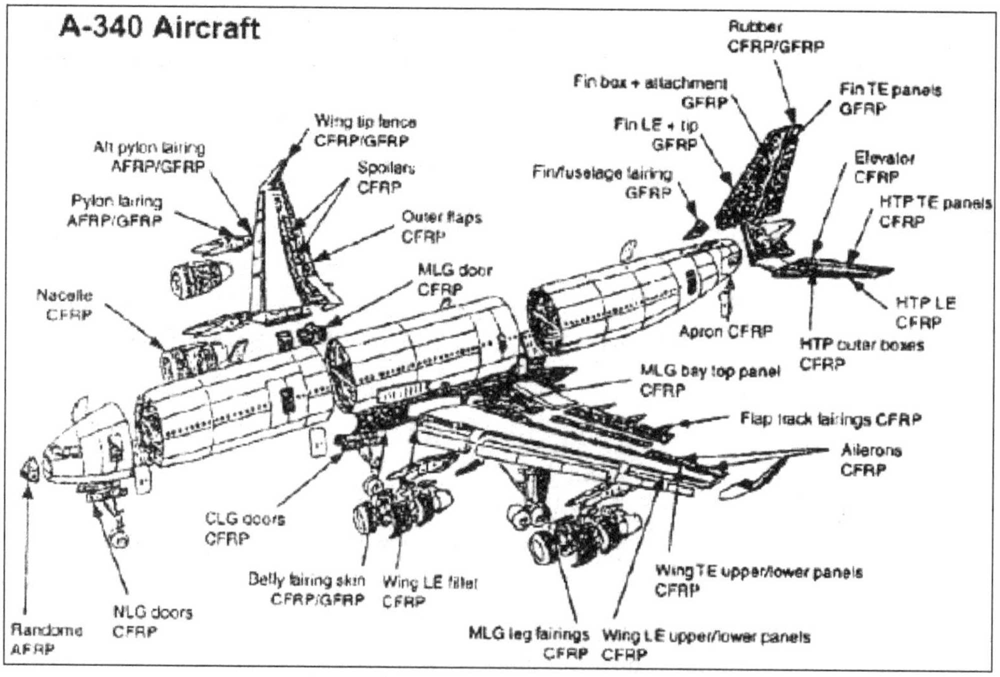

Composite used for different part of A-340 Aircraft.

The first US commercial transport aircraft to fly in 1970, with advanced composite components was Boeing 707 aircraft, which had Boron/Epoxy fore-flap. However, this was only a development production. Commercial aircraft advanced composites development essentially began when Aircraft Energy Efficiency (ACEE) programme was initiated by NASA in 1972. The programme determined properties of different carbon/epoxy material system after long-term worldwide environmental exposure and spectrum fatigue loading equivalent to 20 years or 36,000 flight of airline service. The programme also included flight service of numerous composite components to obtain confidence in the long-term durability of advanced composite structures and materials. The experience gained from the ACEE programme provided the confidence needed by Boeing Co. to select CFRP for the B-757, 8-767 and B-737-300 aircraft control surfaces. Boeing Aircraft Co. designed and 'built Carbon/Epoxy spoilers that were used on Boeing 737-200 aircraft in 1973. The programme included 108 B-737 spoilers, ten B-727 elevators and four B-737 horizontal stabilizers. Some spoilers were later made from different Carbon/thermoplastic materials, which had to be taken out of service due to matrix degradation by hydraulic fluids. Various composite components used on Boeing series of aircraft are given in the table.

Table: The composite components used on Boeing series of aircraft.

Aircraft Type	Components Made of Composite Materials
Boeing 737 -200 -300 -400	Spoilers and horizontal stabilizer (both limited production). Trailing edge flaps. Aileron, elevator, rudder, nacelles. Aileron, elevator, rudder, nacelles.
Boeing 747-400	CFRP winglets and main deck floor panels. CFRP and AFRP used in cabin fittings engine nacelles.
Boeing 757	Aileron, elevator, rudder, spoilers, flaps (in-board & outboard), fairings and nacelles.
Boeing 767	Ailerons, elevator, rudder, spoilers, landing gear doors (nose & main), fairings and nacelles.
Boeing 777	Ailerons, elevator, rudder, spoilers, flaps (in-board & outboard), floor beams, landing gear doors (nose & main), fairings and nacelles.

Use of composites on different versions of new generation B-737 aircraft is shown in figure.

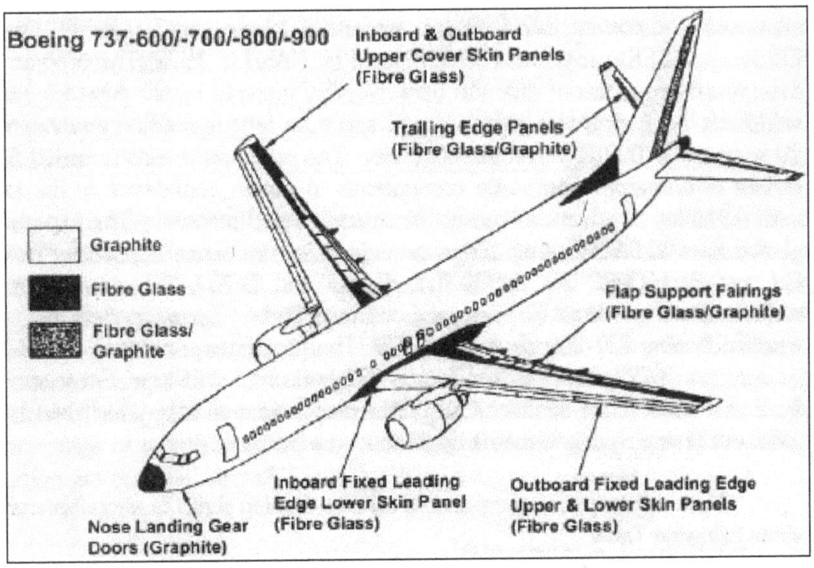

Composite used for different part of B-737 aircraft.

Table: Weight of composites used in the different Boeing series.

Aircraft Type	Weight of Composites
Boeing 737 - 200	≈ 100 lb. Spoiler (1973) ≈ 1150 lb. Horizontal Stabilizer (1984)<1%
Boeing 737 - 300	≈ 1150 lb. (3%)
Boeing 737 - 400	≈ 1150 lb. (3%)
Boeing 757	≈ 3400 lb. (3%)
Boeing 767	≈ 3400 lb. (3%)
Boeing 777	≈ 33000 lb. (10%)

CFRP is used for aileron, elevator, rudder, fairing and engine cowl doors of Boeing 737-300 aircraft. For both Boeing 757 and Boeing 767 aircraft, CFRP composite has been used for elevator, rudders, spoilers, landing gear doors and engine cowlings. The flaps of Boeing 757 aircraft are also made of CFRP. Most of the Boeing 757 and 767 fairings and fixed panels were originally aramid/epoxy or aramid/carbon/ epoxy hybrid-honeycomb sandwich construction. Because of surface and matrix cracking, much of the aramid/epoxy was replaced with glass/epoxy, to improve surface protection.

In case of newly manufactured Boeing 777 aircraft, out of 33000 lb. of composite used, about 10000 lb. is CFRP which includes entire tail, control surfaces, floor beams, main landing gear doors and engine nacelles. The stabilizers have laminated skins with co-cured stringers, solid laminate spars, simple honeycomb sandwich ribs and non-structural forward torque box. They are designed for simple repair. At the moment, Boeing Airplane Co. does not appear to be considering a more extensive use of composites on commercial aircraft. Boeing 777 aircraft, for example, does not use a higher proportion of composites than Airbus A-320 aircraft. The composite structure on B-777 and A-320 aircraft is shown in figures respectively.

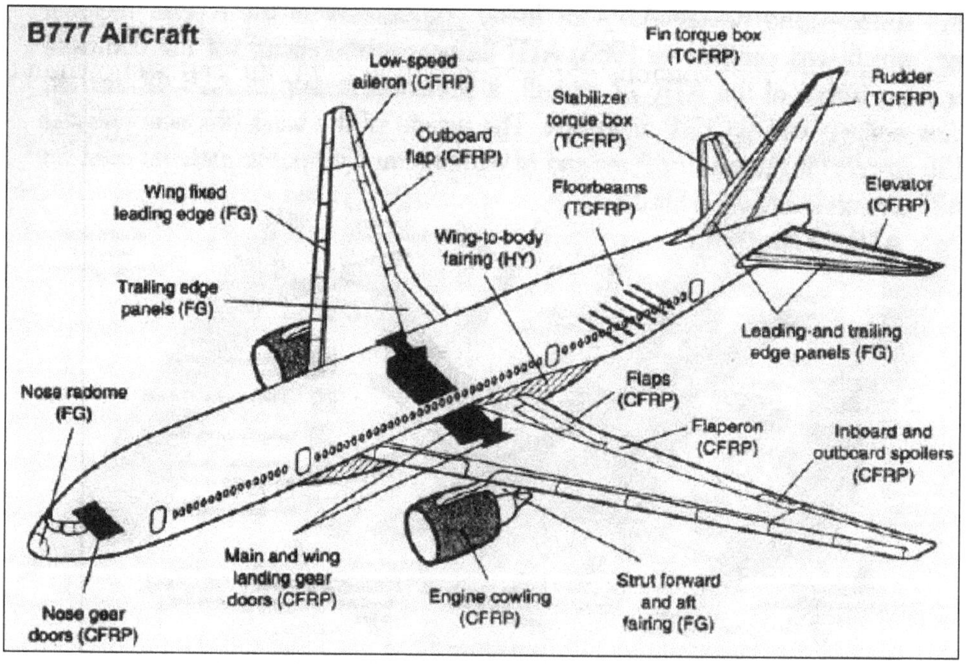

Composite used for different part of B-777 aircraft.

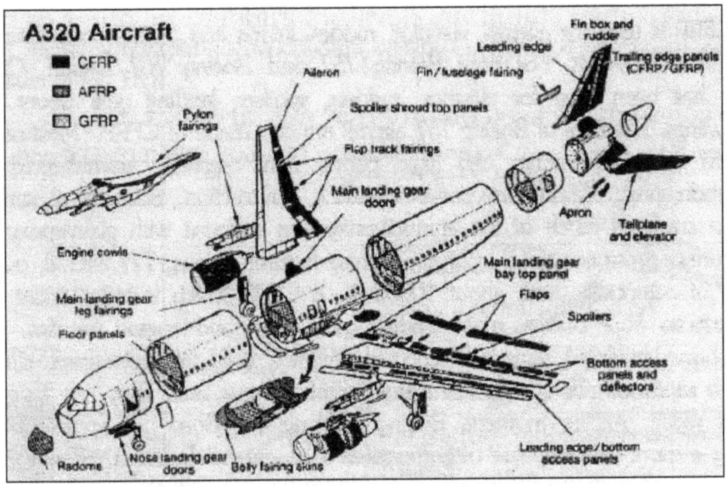

Composite used for different part of A-320 aircraft.

Avions de Transportation Regionale (ATR) uses carbon/epoxy for all the control surfaces and Kevlar/epoxy for many components of the ATR42 regional airline which was certified in 1985. ATR utilizes carbon/epoxy for the complete outer wing boxes of the A TR 72 aircraft, a etched version of A TR 42 aircraft, the first airliner with a CFRP wing box. The weight of the wing box was reduced by 130 kg (286 lb) using CFRP instead of aluminium. Composite material used on ATR-72 aircraft is shown in figure.

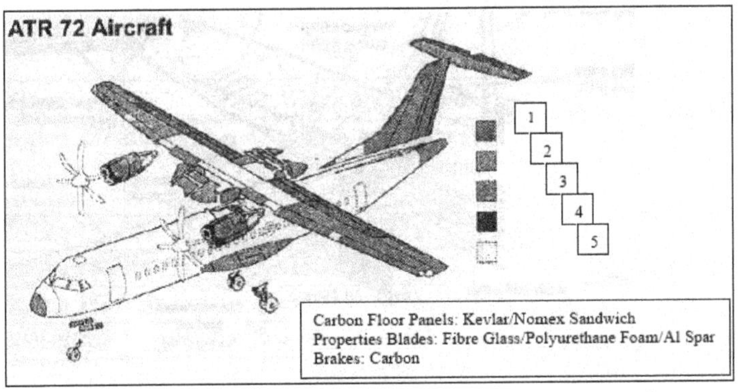

Composite used for different part of ATR-72 aircraft.

In figure, numbers indicates the following:

- 1: Carbon/Nomex sandwich.
- 2: Carbon monolithic structure.
- 3: Kevtar/Nomex sandwich.
- 4: Kevtar/Nomex sandwich with stiffening carbon plies.
- 5: Fibre-glass/Nomex sandwich.

Military Aircraft Applications

Most military aircraft applications use carbon fibre reinforced epoxy composites. About 26% of the structural weight of AV-8B aircraft of carbon fibre reinforced composite comprises the wing box, forward fuselage, horizontal stabilizer, elevators, rudder and other control surfaces. Various composite components used on different military aircraft are given in the table.

Table: The composite components used on different military aircraft.

Aircraft Type	Components Made of Composite Materials
F-14	Doors, Horizontal tail and fairing.
F-15	Rudder, vertical tail, horizontal tail and speed brake.
F-16	Vertical tail and horizontal tail.
F-18	Doors, vertical tail, horizontal tail, wing box, fairings and speed brake.
B-1	Doors, vertical tail, horizontal tail, flaps and slats.
AV-8B	Doors, rudder, vertical tail, horizontal tail, aileron, flaps, wings box, body and fairings.
Typhoon	Wings, fin, rudder, in-board aileron and fuselage.
Light Combat Aircraft (LCA)	Wing, fin, rudder, control surfaces, radome.

On the F-18 aircraft, carbon fibre reinforced composites made up approximately 10% of the structural weight and more than 50% of the surface area.

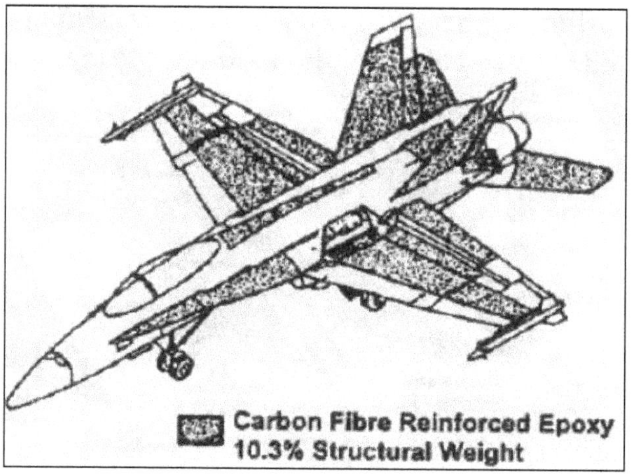

F-18 aircraft.

Similarly, on the Typhoon aircraft, carbon fibre reinforced composites make up approximately 30% of the structural weight and more than 70% of the surface area.

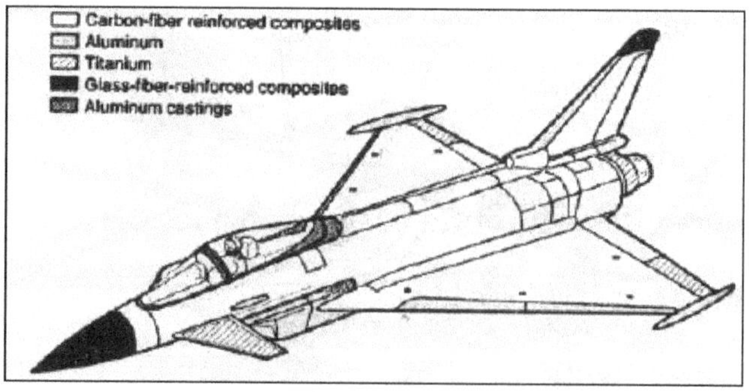

Typhoon aircraft.

The skins of the Typhoon's wing are made of solid laminate, which are shown in figure.

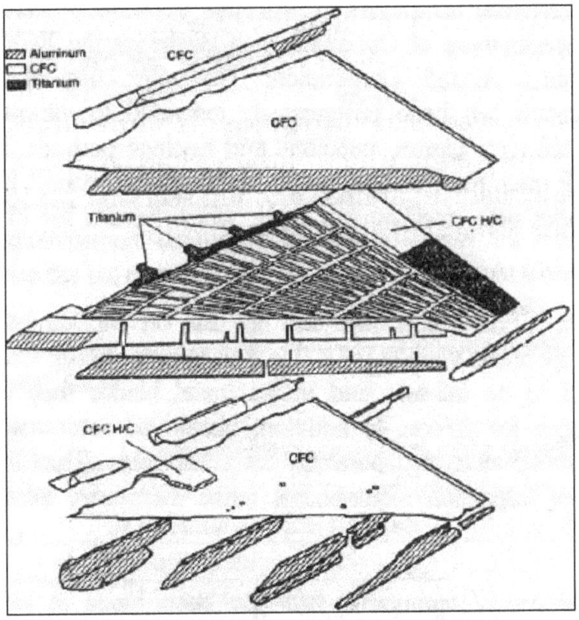

Skins of the Typhoon's wing are made of solid laminate.

On the LCA aircraft, more than 40% of the structural weight of the aircraft is built in advanced composite materials. Entire wing, vertical tail and control surfaces are made of carbon fibre composites.

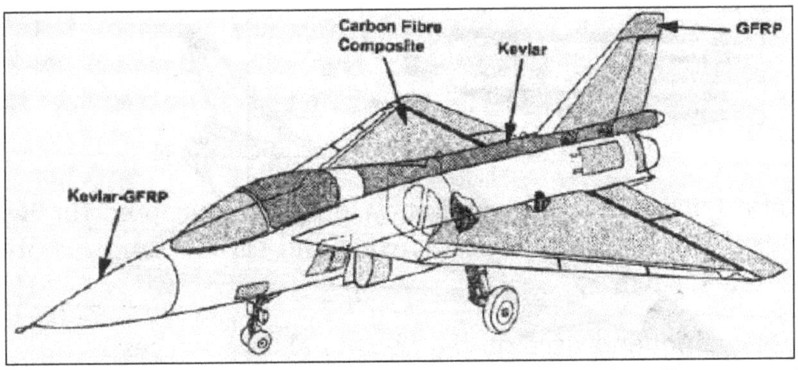

Entire wing, vertical tail and control surfaces are made of carbon fibre composites.

General Aviation Applications

Composite materials are being used for different helicopter components as well. Use of advanced composites in helicopter application started way back in 1959 with the development of Optimum Pitch Blade for the XCH-47 twin rotor helicopter of Vertol Aircraft Corporation. There-after, use of composites in helicopter application has been progressively extended to various parts, which include main & tail rotor blades, stabilizers and fuselage portions. Experience has shown that GFRP main rotor blades have a service life of around 10,000 hours as compared to blades with steel/ titanium spars, which have a life of around 1000 - 2000 hours.

Flex-beam CFRP rotors are also in use on McDonnell Douglas MD 520N/MD 900 helicopters with NOTAH (No Tail Rotor) system. Composite blades can be designed to be fail-safe and unlike metal blades they do not require frequent inspections for defects. In addition, blade and rotor system efficiencies have been improved due to tailorability of composites. Their longer life and reduced in-service inspection requirements make them very attractive and cost effective.

Extensive use of composites has also been made in India's Advanced Light Helicopter (ALH). In ALH composite material is employed in whole of the secondary structure and several parts of the primary structure. The nose is made of aramid and tail section of carbon fibre reinforced plastic. Rotor hub, main and tail rotor blades are made of composites. Entire cockpit is also made of composite material. The extent of composites used in the structure is about 60% by wetted area and about 29% by weight.

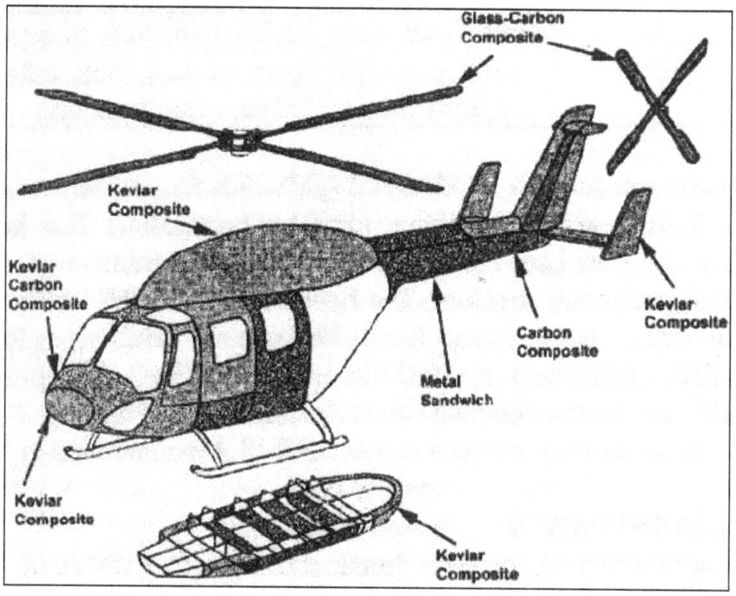

The Cockpit is also made of composite material.

Eurocopter has used extensive composites on its latest civil helicopter, the EC 135. Using glass/carbon composite for the fenestron tail rotor shroud and carbon sandwich for the tail cone has yielded appreciable weight savings.

Table: The composite components used on significant helicopter types.

Helicopter Type	Components Made of Composite Materials
MBB BK 117	Main rotor blades, tail rotor blades, horizontal stabilizer, vertical stabilizer.
Bell 206L	Vertical stabilizer.
Bell 402	Main rotor blades.
Dauphin	Main rotor blades, vertical stabilizer.
McDonnell Douglas MD 520N	Main rotor blades, tail boom.
McDonnell Douglas MD 900	Main rotor blades, fuselage mid section, tail boom, canopy frame, internal fuselage, horizontal stabilizer, vertical stabilizer.
Advanced Light Helicopter (ALH)	Main & tail rotor blades, rotor hub, nose cone, crew & passenger doors, cowling, most of the tail unit, lower rear tail boom, cock it section.

In general aviation, the Windecker Eagle, which flew in 1967, was the first all composite GFRP airplane to obtain full FAA certification. The Eagle was followed by the Lan Avia Lear Fan 2100, which had an airframe made of CFRP with some AFRP secondary structure. The Beech Starship 2000, which is almost entirely made from carbon/epoxy faced honeycomb sandwich construction, received full FM certification in 1988. HANSA-3, an all-composite trainer aircraft has been designed and developed by National Aerospace Laboratories. The aircraft was certified to meet JAR-VLA requirements in 1999.

Structural Components

Composite structural and semi-structural components, which are currently being produced, are given in the table.

Table: Composite structural and semi-structural components.

Component	Details
Wing	• Box beam skins • Box beam sub-structure • Winglets • Leading edge flaps/slats • Ailerons/flaperons • Raps & Spoilers • Fixed leading edges • Fixed trailing edge panels • Rap track fairings • Actuator fairings
Empennage	• Horizontal stabilizers • Skins • Sub-structure • Elevators • Leading edges • Fixed trailing edge panels • Tips
Vertical Stabilizers	• Skins • Sub-structure • Rudders • Leading edges • Fixed trailing edge panels • Ventral fins • Tips

Fuselage	• Radome • Forward fuselage • Canopy frames (helicopters) • Mid fuselage • Rear fuselage • Speed brakes • Tail cone • Floor beams • Floors • Rotor-domes • Cabin doors (helicopters) • Lining and partitions • Overhead baggage compartment • Air ducts
Helicopter	• Main rotor blades • Tail rotor blades • Rotor drive shafts
Doors and Fairings	• Landing gear doors • Landing gear fairings • Landing gear pods • Wing-fuselage fairings • Stabilizer fairings • Equipment access doors
Propulsion System	• Engine fan blades • Engine casing • Nozzle flaps • Thrust reversers • Engine nacelle and cowling • Fan cowls • Turbine blade containment rings • Pylon fairings • Fuel tanks • Propeller blades

Composites for Structural Applications

Composites have long been used in the construction industry. Applications range from non-structural gratings and claddings to full structural systems for industrial supports, buildings, long span roof structures, tanks, bridge components and complete bridge systems. Their benefits of corrosion resistance and low weight have proven attractive in many low stress applications. An extension to the use of high performance FRP in primary structural applications, however, has been slower to gain acceptance although there is much development activity. Composites present immense opportunities to play increasing role as an alternate material to replace timber, steel, aluminium and concrete in buildings.

Construction

Construction holds priority for the adaptation of composites in place of conventional materials being used like doors and windows, paneling, furniture, non-structural gratings, long span roof structures, tanks, bridge components and complete bridge systems and other interiors. Components made of composite materials find extensive applications in shuttering supports, special architectural structures imparting aesthetic appearance, large signages etc. with the advantages like corrosion resistance, longer life, low maintenance, ease in workability, fire retardancy etc. Some of the composite structural applications are listed in the table.

Table: Composite applications in building & construction.

Composite	Applications
Coconut/jute/sisal fibre and gypsum plaster	Boards for partitions, ceiling and wall panel.
Coconut/jute/sisal fibre and glass fibre with gypsum plaster	Solid and perforated building blocks.
Poplar wood with polymer	Door shutters, doors and window frames.
Vegetable fibre/red mud and polyester	Flat and corrugated roofing sheets, shutters and tiles.
Jute fibre polyester/epoxy with red mud	Panels and sheets, wall cladding, partitions and door shutters.
Bagasse & UF/PF resin	Panels and blocks.
Coir fibre with fly ash & lime or cement	Bricks & blocks for walling.
Glass fibre reinforced polyester	Roofing sheets, partitions, ceilings, sanitary ware, water storage tanks, pipes etc.
Glass fibre reinforced PP/nylon	Door hinges.
Red mud plastic (PVC, polyester) reinforced with sisal fibre	Paneling, roofing, partitions, door panels.
Acrylic resin and quartz sand composite	Kitchen sinks.

Usage of composites for damage repairing, seismic retrofitting and upgrading of concrete bridges finds increased adoption as a way to extend the service life of existing structures, they are also being considered as an economic solution for new bridge structures. Composite based 2D and 3D grid-type reinforcement for concrete structures shows considerable potential for use as reinforcement of concrete in tunnels because of its corrosion and chemical resistance, its lightweight and its ease of forming to fit curvatures. Grid-type reinforcement for concrete structures consists of high-performance fibres such as glass, carbon, aramid and hybrids impregnated with resin systems ranging from vinyl esters and other thermosetting resin systems to thermoplastics. The significant systems-level savings were achieved due to the factors of weight. Other critical applications of composites in the civil engineering area are:

- Tunnel supports.
- Supports for storage containers.
- Airport facilities such as runways and aprons.
- Roads and bridge structures.
- Marine and offshore structures.
- Concrete slabs.
- Power plant facilities.
- Architectural features and structures such as exterior walls, handrails, etc.

The key restricting factors in the application of composites are initial costs due to raw materials and also inefficient moulding processes. With the adoption of advanced technologies and some extent of standardization, these problems could be easily taken care of. A growth rate of 11-13% p.a. in the usage of composites is expected after 2005 AD in this sector.

Road Bridges

Bridges account for a major sector of the construction industry and have attracted strong interest for the utilization of high performance FRP. FRP has been found quite suitable for repair, seismic retrofitting and upgrading of concrete bridges as a way to extend the service life of existing structures. FRP is also being considered as an economic solution for new bridge structures. Polymer composites are seen to offer advantages that are lacking in the traditional materials, particularly for their resistance to corrosive attack in those areas that rely on the application of de-icing salts to maintain road access. Design approaches and manufacturing efficiencies developed for road bridge applications will benefit their introduction into a broader range of civil construction fields.

The road bridge.

Decks for both pedestrian and vehicle bridges across waterways, railways and roadways are now a commercial reality; with some pedestrian bridges being built entirely from composites. The lightweight of composites is especially valuable for the construction of waterway bridges incorporating a lift-up section to permit the passage of boats, and for ease of transportation and erection in remote areas without access to heavy lifting equipment. The composite deck has six to seven times the load capacity of a reinforced concrete deck with only 20 percent of the weight. Because of the superior durability of composite, only cosmetic maintenance requirements are expected for at least 50 years. The composite will not spall like concrete during freeze-thaw cycles and will not rust like metal in the moist, corrosive sea environment. The composite bridge decks are quite suitable for replacing conventional/old bridge decks having super structure intact. The replacement can be carried out in a short time with minimal disturbance to the traffic.

The ability to use composites to design a lightweight, prefabricated module brings about immediate cost savings by minimizing the disruption of traffic and commerce. The lightweight composite module can be installed in a matter of hours instead of the days or weeks it takes to replace a deteriorated bridge deck with a conventional one. The high strength-to-weight ratio of the composite deck reduces the costs associated with heavy equipment. Lower deck weight also offers the potential for a higher load rating. Composites can significantly reduce maintenance and replacement costs because of the material's excellent resistance to corrosion and fatigue. In essence, composite durability not only improves life-cycle costs but extends the life-cycle itself.

The bridge deck sections, composed of hexagon and double-trapezoid profiles are bonded with a high-strength adhesive under controlled conditions. The prefabricated sections are then transported to the job site, ready for installation. Composite bridge decks replace deteriorated concrete or timber bridge decks and their lighter weight permits easier & quicker installation. FRP composite bridge decks are made of pultruded components that are bonded and interlocked. They are placed transversely to the traffic and are supported by longitudinal beams. The FRP decks comprise of double trapezoid composite connected with full depth hexagons that provide mechanical interlock and an extensive bonding surface. Construction of highway bridges with modular FRP decks requires the understanding of the deck performance under traffic loads. Traffic loads induce repetitive stress cycles on bridge decks during the service life of the structure. The composite bridge decks are modular in design and can be produced in continuous lengths because of the inherent process adopted (pultrusion technique) and these lengths can be cut to size depending on the user's requirement. Hence, it furnishes greater flexibility in fabrication of the composite bridge decks to suit various product dimensions.

The first FRP modular bridge decks in USA were installed in rural roads in West Virginia. Composite bridge decks performed satisfactorily for 2 million cycle loads without major fatigue. In addition to the deck, the Laurel Lick installation uses composite, stringers, piling and short-column abutments. The decks exhibited a safe failure mode with considerable energy absorption and adequate post failure reserve strength. The installations in West Virginia demonstrate how the performance characteristics of fiber-reinforced polymer (FRP) composites are ideally suited for highway bridge construction. This has been a significant step in the development of modular bridge decks made with standardized "off-the-shelf" components.

The 40-m (131-ft.) long, 3-m (9.8-ft.) wide crossing carries pedestrians, bicycles and motorbikes over a previously dangerous set of railroad tracks. As the bridge was designed to support a 500 kg/m^2 (102 lb. /ft.^{2}) load, the structure can also Composite Bridge Decks accommodate snow removal vehicles weighing up to 5 MT. This impressive strength is provided by a composite deck that weighs only 12 MT. In steel and concrete, a bridge deck of comparable strength would weigh 28 MT and 90 MT respectively. The bridge is suspended from a high support tower (18.5-m (61-ft.)) that is bolted to a concrete foundation. The composite tower weighs only 3 MT. Fiber-line Composites used its pultruded profiles to pre-fabricate the tower and three bridge sections for final assembly at the bridge site. The lightweight composite allowed the bridge to be easily erected in only 18 night-time hours, thus minimizing disruption to rail traffic.

Bridge Structures

Composite bridge decks.

E.T. Techtonics Inc., Philadelphia, USA, designs Prestek truss and pre-stressed cable bridges that are assembled of structural composite. The components are standard pultruded profiles. The Prestek design is especially attractive for areas where the impact of construction must be minimized. The bridges' pultruded composite components deliver critical mechanical properties at a much lower weight than steel. As a result, a Prestek span is easily installed without heavy equipment or large work crews. Prestek bridges have been specified for such exotic locations as a national park on the island of Maui and a ridgeline leading to a historic lighthouse off the Golden Gate. Typical composite bridge decks as shown in figure.

Pultruded Profiles

Among a wide array of composite products, pultruded profiles such as gratings, ladders, cable trays, solid rods & other sections are used in many structural application with Class I flame retardancy. Pultrusion is the most cost-effective method for the production of fibre-reinforced composite structural profiles. It brings high performance composites down to commercial applications such as lightweight corrosion-free structures, electrical non-conductive Pultruded FRP Sections systems, offshore platforms and many other innovative new products. The pultruded profile is shown in figure.

Pultruded sections are well-established alternative to steel, wood and aluminium in developed countries and are fast catching up in other parts of the world. Structural sections have ready markets in oil exploration rigs, chemical industries etc. The amount of energy required to fabricate FRP composite materials for structural applications with respect to conventional materials such as steel and aluminium is lower and would work for its economic advantage in the end.

The pultruded products are already being recognized as commodity in the international market for construction.

The pultruded profile.

In pursuit of developing advanced performance materials for building and construction, railways, automobiles, bio-medical etc., the Advanced Composites Programme was launched by Technology Information, Forecasting and Assessment Council (TIFAC), an autonomous organization under the Department of Science and Technology (DST), Govt. of India. Under a project of the afore-said programme, FRP Pultruded profiles (industrial gratings, solid rods for electrical insulation, cable-trays, ladders etc.) with excellent surface finish and flame retardancy as per international standards have been developed. The profiles developed have met all the desired properties. Table lists the mechanical/chemical properties of FRP pultruded sections vs. other structural materials. Table lists the advantages of pultruded products. Table lists out the characteristics of the pultruded products.

Under a project of the Advanced Composites Mission programme of the Govt. of India, FRP Pultruded profiles (industrial gratings, solid rods for electrical insulation, cable-trays, ladders etc.) with excellent surface finish and flame retardancy as per international standards have been developed. Orders were received from various industries for supplying cable trays, fittings and other accessories. An order was also executed for supplying cable trays, fittings and other accessories to Company A as per the specifications of Company B. The profiles developed have met all the desired properties.

The pultruded profile.

Table: Mechanical/chemical properties of FRP pultruded sections vs. other structural materials.

Mechanical Properties	Pultruded FRP		Rigid PVC	Mild Steel	Stainless Steel	Wood
	Polyester	Vinyl Ester				
Tensile Strength (N/mm^2)	382	401	44	340	340	80
Flexural Strength (N/mm^2)	468	508	70	380	380	12
Flexural Modulus (N/mm^2)	22489	48260	2400	196000	196000	700
Izod Impact (Kg-m/cm)	1.36	1.63	0.09	1.5	0.53	---
Specific Gravity	1.80	1.80	1.38	7.8	7.92	0.52
Safe Working Temp. (°C)	120	170	55	600	600	160

Table: Advantages of pultrusion.

Features	Description	Benefits	Applications
Strong	Unit strength intension & compression is approx. 20 × that of steel when these properties are combined on the basis of unit density.	Optional strength as desired. Exceptionally high impact strength reduces damage potential.	Structural process equipment support. Tank supports. Cooling tower ancillaries. Flooring supports. Trusses & joints.
Light Weight	Density of pultruded components is about 20% of steel and 60% of aluminium.	Higher performance at less weight. Lower shipping, handling & installation costs. Less operational energy demand.	Automotive leaf springs & bumpers. Prefabricated walkways & platforms. Bus components.
Corrosion Resistant	Unaffected by exposure to a great variety of corrosive environment & chemicals.	Minimum maintenance costs. Long term safety. Longer life.	Chemical plant hand railings, gratings, walkways & bridges. Cable trays. Pipe supports.
Electrical Insulation	Provides strength & rigidity with dielectric properties.	Lesser no. of components. Non-magnetic & safe. Predictable insulation values for wide range of frequencies.	Ladders, Cable trays. Switch gear components. Mounting braces and backboards.
Thermal Insulation	Pultruded components have a low thermal conductivity, 1/250 of aluminium & 1/60 of steel.	Reduces installation thickness. Eliminates condensation problems. Reduces energy operation requirements.	Bulk head frames. Walk in refrigerator door jams. Window frames. Insulated roll up panel doors.
Consolidation	Many individual components can be combined into a large profile.	Reduced assembly cost. Reduced inventory. Fewer parts improve reliability.	Window latch supports. Roll up door
Dimensional Stability	Pultruded components are highly resistant to warping stretch/swelling over a wide range of temperature & stresses.	No permanent deformation under high stress. Close tolerances.	Spring bumpers. Crossing gate arms. Scrubber components.
Safety	The pultruded components are very strong & safe to work with. They are microbes and insect proof.	Many gratings suffer from the problem of microbes etc. due to wet or unhygienic working conditions.	This property makes them ideal choice for pharmaceutical & food industries.

Table: Pultruded product characteristics.

Size	Forming guide system and equipment pulling capacity influence size limitation.
Shape	Straight, constant cross sections, some curved sections possible.
Length	No limit.
Reinforcement	Fibre glass, aramid fibre, carbon fibre, thermoplastic and natural fibres.
Mechanical Strength	Medium to high, primarily unidirectional approaching isotropic.
Labor intensity	Low to medium.
Mould cost	Low to medium.
Production rate	Shape and thickness related.

The profiles like gratings, solid rods for electrical insulation, cable trays, ladders etc. were developed successfully with technology support from National Chemical Laboratory (NCL), Pune with excellent surface finish and flame retardancy as per international standards. NCL had worked & established various critical parameters such as using 3-cataylst system for improved curing, high pultrusion speed (1.0 – 1.5 m/min) and flame retardancy characteristics for the products. The comparison chart of the properties of FRP pultruded sections and other structural materials are listed in table.

Table: Physical & chemical properties of pultruded profiles vs. other structural materials.

Properties	Pultruded FRP	Rigid PVC	Mild Steel	Stainless Steel	Wood
Specific Gravity	1.8	1.38	7.8	7.92	0.52
Thermal Conductivity (Kcal/hr/m²/°C)	24.4	6.4	1220	732.00	0.4
Coeff. of Linear Expansion (cm/cm° C) × 10-6	5.2	37	8	10	1.7
Safe Working Temp. (°C)	130	55	600	600	160
Flame Resistance	Good*	Poor	Excellent	Excellent	Poor
Corrosion Resistance					
a. Acidic	Excellent	Good	Poor	Excellent	Poor
b. Alkaline	Good	Fair	Good	Excellent	Poor
c. Solvents	Fair	Poor	Good	Excellent	Fair
d. Coastal Environment	Excellent	Good	Poor	Excellent	Fair
e. Outdoor Exposure	Excellent	Poor	Fair	Excellent	Fair
f. Effluent Water	Excellent	Good	Poor	Excellent	Fair
g. Steam	Good	Poor	Fair	Excellent	Fair
* Excellent with special additives					

Towards market seeding of the pultruded products, the Company targeted three major segments viz. new projects, replacement market in industrial and non-industrial applications. The products such as cable trays, fittings and other accessories are being inducted by various industries and abroad.

Power Transmission

High voltage electrical transmission towers are now being constructed from pultruded composite sections using a "snap and build" assembly procedure, which eliminates the use of fasteners and adhesives. Weighing less than one third of conventional steel equivalent structures, the composites tower components can be readily airlifted into remote areas and assembled by small teams, thus eliminating the need to construct access roads. In addition, the inherent insulating characteristics of composite permits closer placement of attached insulators, thus enabling overall tower size and environmental obtrusiveness to be reduced. In coastal environments, the corrosion resistance of composite towers offers a huge advantage. Test towers located at California's Pacific coastline have now been in service for over three years and have withstood heavy salt contamination without development of arcing which results in insulator & consequent system failure. By contrast, galvanized steel towers require frequent washings to remove conductive, corrosion induced salt deposits.

Power Distribution and Lighting

Composite power and lighting poles are finding increased application for both performance and environmental reasons. Traditional wooden poles contribute to forest depletion. In order to protect them against rot and termite attack, they also require treatment with highly toxic chemicals, which have been found to leach out into the surrounding environment. Lightweight composite poles are easily transported and erected; they resist corrosion, rot and insect attack, provide superior insulation properties, and can be designed to substantially reduce the possibility of collision fatalities when close placement to roads is unavoidable.

Repair, Retrofit and Rebars

Composite plates are successfully used to repair masonry beams, columns, buildings and other structures damaged/weakened by impact, earthquake or subsidence and can usually be adhered in place by hand without the need for heavy lifting equipment. Such repairs can be carried out much more rapidly than by traditional techniques. Composite reinforcing bars may be used to replace steel in conventional reinforced concrete in order to prevent "concrete cancer" problems resulting from internal corrosion of the reinforcement. Although initially costlier, the use of composite rebars is justified where the nature of the construction would render possible future repairs inaccessible or otherwise unduly costly.

FRP Doors and Door Frames

With the scarcity of wood for building products, the alternative, which merits attention, is to promote the manufacturing of low cost FRP building materials to meet the demands of the housing and building sectors. The doors made of FRP skins, sandwiched with core materials such as rigid polyurethane foam, expanded polystyrene, paper honeycomb; jute/coir felt etc. can have potential usage in residential buildings, offices, schools, hospitals, laboratories etc. As structural sandwich construction has attained broad acceptance and usage for primary load bearing structures, the FRP doors can be manufactured in various sizes and designs using this technology.

The principal fabrication technique employed is contact moulding or hand lay-up process. The front and back sheets of the doors are fabricated separately. Wooden inserts are placed between two sheets for various fittings. The PU foam is sandwiched between the sheets by in-situ foaming process followed by painting and polishing to meet aesthetic requirement. Proper usage of additives imparts fire retardant properties to the doors. In addition, usage of composite material for the doors makes them totally water & termite resistant. FRP doors are much cheaper than the wooden ones. The FRP doorframes can also be fabricated by contact moulding. The FRP doors and doorframes have been designed & developed using the aforesaid technology by the RV-TI-FAC Composite Design Centre (CDC) at Bangalore under the Advanced Composites Mission programme of the Govt. of India. The FRP doors developed by CDC conform to BIS specifications (IS: 4020). After successful field trials and users' feedback, the technology for FRP door has been transferred to over 30 entrepreneurs for commercial exploitation.

The rapid expansion of the use of sandwich construction in many fields has yielded a more precise knowledge of design methods, test procedures and manufacturing techniques of cost-effective composite products. A low-density core made of honeycomb or foam materials provides a structural performance with minimum weight. Other considerations such as sound insulation, heat resistance, vibration-damping etc. dictate the particular choice of material used as core material. The FRP door is shown in figure.

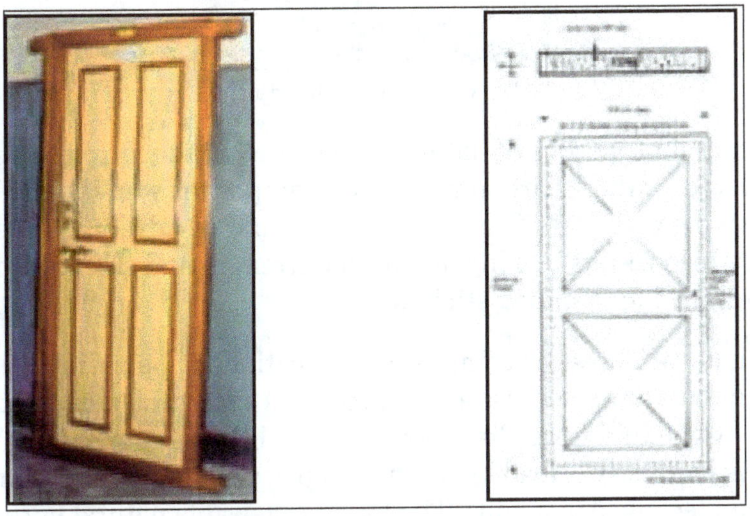

FRP Door & Door Frame.

- Plumbing Components: Lightweight fibre glass composite components for toilet are easy to install and they are corrosion resistant. Due to poor thermal conductivity, the composite surface is warm to the touch unlike porcelain and steel. Ease in moulding technique for composite allows more aesthetic shapes and excellent surface finishes.
- Ceiling Panel: The fibre glass veil facing used while moulding the panels for suspended ceilings increases panel stiffness and resists puncturing. Due to their easy printability, the veil imparts good panel aesthetics. The suspended ceilings are used to cover up electrical wiring, ducting, piping and fittings. The veil with an optimum porosity contributes to improved acoustical quality of the working or living space.

Natural Fibre Composites in Building Materials

Natural fibres, as a substitute for glass fibres in composite components, have gained interest in the last Composite Ceiling Panel decade, especially in the housing sector. Fibres like flax, hemp or jute are cheap, have better stiffness per unit weight and have a lower impact on the environment.

Structural applications are rare since existing production techniques are not applicable and availability of semi-finished materials with constant quality is still a problem. Typical composite material is shown in figure.

Typical composite building.

The moderate mechanical properties of natural fibres prevent them from being used in high-performance applications (e.g. where carbon reinforced composites would be utilized), but for many reasons they can compete with glass fibres. Advantages and disadvantages determine the choice. Low specific weight, which results in a higher specific strength and stiffness than glass is a benefit especially in parts designed for bending stiffness.

Recently the use of natural fibres for composite applications is being investigated intensively in Europe. As a result, many components are now produced in natural composites, mainly based on polyester or polypropylene and fibres like flax, jute, sisal, banana or ramie. Until now however, the introduction in this industry is lead by motives of price and marketing ('processing renewable resources') rather than technical demands. The use of renewable natural fibres contributes to sustainable developments.

Natural fibre composites (NFC) can be used as a substitute for timber as well as for a number of other applications. It can be moulded into sheets, boards, gratings, pallets, frames, structural sections and many other shapes. They can be used as a substitute for wood, metal or masonry for partitions, false ceilings, facades, barricades, fences, railings, flooring, roofing, wall tiles etc. It can also be used in pre-fabricated housing, cubicles, kiosks, awnings, sheds/ shelters. Natural fibres due to their adequate tensile strength and good specific modulus enjoy the right potential for usage in composites thus ensuring a value-added application avenue. The maximum tensile, impact and flexural strengths for natural fibre composites reported so far are 104.0 MN/m^2 (jute-epoxy), 22.0 kJ/m^2 (jutepolyester) and 64.0 MN/m^2 (banana-polyester) respectively.

Although the tensile strength and Young's modulus of natural fibre like jute are lower than those of glass fibres, the specific modulus of jute fibre is superior to that of glass and on a modulus per cost basis, jute is far superior. The specific strength per unit cost of jute, too, approaches that of glass. The need for using jute fibres in place of the traditional glass fibre partly or fully as reinforcing agents in composites stems from its lower specific gravity (1.29) and higher specific modulus (40 GPa) of jute compared with those of glass (2.5 and 30 GPa respectively). Apart from much lower cost and renewable nature of jute, much lower energy requirement for the production of jute (only 2% of that for glass) makes it attractive as a reinforcing fibre in composites. The table shows the properties of natural fibres.

Table: Properties of select natural fibres.

Properties	E-glass	Flax	Hemp	Jute	Ramie	Coir	Sisal	Cotton
Density g/cm^3	2.55	1.4	1.48	1.46	1.5	1.25	1.33	1.51
Tensile Strength * 10E^6 N/m^2	2400	800-1500	550-900	400-800	500	220	600-700	400
E-modulus (GPa)	73	60-80	70	10-30	44	6	38	12
Elongation at Failure (%)	3	1.2-1.6	1.6	1.8	2	15-25	2-3	3-10
Moisture Absorption (%)	---	7	8	12	12-17	10	11	8-25
Price / kg ($), raw	1.3	0.5-1.8	0.6-1.8	0.35	1.5-2.5	0.25-0.5	0.6-0.7	1.5-2.2
* Tensile strength strongly depends on type of fibre, being a bundle or a single filament.								

Jute-Coir Composites

An alternative to wood products is one of the application avenues for construction industry. It involves the production of coir-ply boards with oriented jute as face veneer and coir plus waste rubber wood inside. A very thin layer of jute fibres impregnated with phenolic resin is used as the face veneer for improved aesthetics and to give a wood like finish. The orientation & uniformity of jute fibre improve with carding and this also helps in better penetration of resin into the fibre. The coir fibre contains 45.84% lignin as against 39% in teakwood. Therefore, it is more resistant than teakwood against rotting under wet and dry conditions and has better tensile strength. Similarly low cellulose content in coir (43%) as against 63% cellulose in wood makes it more durable than teakwood. The composite boards namely, coir-ply boards (jute + rubber wood + coir) as plywood substitute and natural fibre reinforced boards (jute + coir) as MDF substitute can be used in place of wood or MDF boards for partitioning, false ceiling, surface paneling, roofing, furniture, cupboards, wardrobes etc. These boards have been employed as doors & door frames as an alternate to conventional material like wood, steel etc. The properties of jute-coir boards developed by the 'Natura' Division of company under the Advanced Composites Mission programme were tested as per IS-12406 against the specified values of MDF a board is given in table.

Table: Properties of jute-coir boards tested as per IS-12406.

Sl. No.	Tests	Observed Values(Average)			Specified Values
		Board Thickness			
		8 mm	6 mm	4 mm	
1.	Cross Breaking Strength (Kgs./cm^2) -Perpendicular to Grain Direction)				
(a)	Before Boiling	318	391	373	275 (min.)

(b)	After 8 Hrs. Boiling	266	270	240	150 (min.)
2.	Bulk Density (Kgs./cm³)	700	739	760	500-900

Table: Properties of jute-coir boards tested as per IS-12406.

l. No.	Tests	Observed Values (Avg.)			Specified Values	
		Board Thickness				
		8 mm	6 mm	4 mm	Exterior Grade	Interior Grade
1. (a)	Moisture Content (%)	5.73	5.90	5.92	5-15	5-15
(b)	Variation from mean moisture content (%)	− 2.1	+ 0.9	+ 1.2	+3.0	+ 3.0
2.	Max. Water absorption (%)					
	After 2 Hrs. soaking	4.5	5.1	2.9	6	9
(a)	After 24 Hrs. soaking	9.1	9.2	6.8	12	18
3.	Max. linear expansion (% swelling in water)					
	Due to general absorption after 24 Hrs. soaking					
	Thickness	Average value : 1.0			4	7
	Length	Average value : 0.13			0.3	0.4
	Width	Average value : 0.21			0.3	0.4

The bamboo based composites can be used in construction sector for blocks, flooring tiles, false ceilings, partition walls, furniture etc. with laminated or polymerized surface. Bamboo after treatment with chemicals for termite and fungi is stored in solar chamber for drying. Bamboo poles are cut manually to sizes. The slats are measured using different gauges to arrive at proper sizes. Bamboo slats are impregnated with polyester resin and moulded in the hydraulic press. The moulded product is machined for exact dimensions and lacquer coated for finishing. The bamboo based composite furniture as shown in figure.

The bamboo based composites furniture.

Due to the easiness to drill, tap, shape and assemble these blocks are used for making fixtures tools handles sports goods, etc. Pre-fabricated housing elements for sheds, shelters, cubicles, huts, pyramids etc. are also fabricated by using bamboo composites. These elements comprise of structural frames, beams, columns, trusses, rafters etc. and are erected by cladding with NFC sheets. They are ideal for remote and inaccessible laces and very much suitable for hill forts, isolated beaches, inclement weather stations, conventional as well as unconventional tourist spots, farmhouses, retreats and small islands. Other product that is developed using bamboo is decorative board. It is made by sized flats of bamboo having matching thickness, which are then joined / bonded together by pouring polymer resin. Due to the inherent lightness of the bamboo the decorative boards that are developed also gain this feature of bamboo, therefore it becomes lighter than the other wood and plywood boards.

Bamboo Composite Boards and Laminates

Bamboo is one of the fastest renewable plants with a maturity cycle of 3-4 years, thus making it a highly attractive natural resource compared to forest hardwoods. Bamboo offers good potential for processing it into composites as a wood substitute. Bamboo laminates could replace timber in many applications such as furniture, doors and windows and their frames, partitions, wardrobes, cabinets, flooring etc.

Bamboo laminates are made from slivers milled out from the bamboo culm. After primary processing comprising cross cutting, splitting and 2-side planing, the slivers are treated for starch removal and prevention of termite/borer attack. The slivers are then subjected to hot air drying followed by 4-side planing for attaining uniform thickness. These slivers are coated with glue on the surface and are arranged systematically. They are subjected to a curing in a hot press (6'×4' 2-day light) at temperature ~70 °C using steam & pressure ~17 Kg/cm^2. The pressed laminate (panels/tiles) is then put through trimming, sanding and grooving machines to give a pre-finish shape. The flow chart & intermediate quality control parameters for manufacturing bamboo composites are enclosed. The project on production of bamboo composites and laminates is based on the following premises:

- Value-added products from Bamboo.
- Cost-effective compared to good solid wood sections for furniture.
- Diversification from traditional plywood to bamboo based products.
- Complete range of bamboo composite laminates for furniture, flooring tiles, boards, door & window frames to replace the use of timber for domestic as well as international market.

Expected Benefits

- Bamboo composite based flooring tiles, boards (used for partitions, cupboards, racks, and door and window panels) and blocks (used for furniture, rails and styles for doors and windows etc.) as wood substitute would help develop and promote high value-added products from bamboo.
- Bamboo composite laminates with a low-temperature curing resin system for reduced energy requirement.
- Promotion of eco-friendly use of bamboo while building a sustainable infrastructure for plant multiplication, propagation and cultivation.

- Boosting the usage of bamboo based products towards generating good employment and income opportunities at rural level.

Towards effective bamboo utilization and exploring the value-addition potential, the project on development of bamboo composite laminates was launched by the Advanced Composites Programme of TIFAC in partnership with companies with technology support from the Department of Polymer Science and Technology, University of Calcutta. The project aimed at developing value-added products from bamboo with an innovative resin system for reduced processing energy requirement. Bamboo based products such as flooring tiles, laminate boards; blocks (for door and window frames, rails and styles, furniture etc.) as wood substitute are being developed under the project.

For preventing bamboo composites from any deterioration by moisture absorption and imparting long-term storage life, a water based acrylic pre-coat has been developed. This pre-coat would prevent any fungal attack during transit for the reconstituted wood sections for furniture. Further, a UV cured melamine acrylate system as the finishing coat has also been developed for flooring tiles made of bamboo composites. A water based PU resin system has also been tried for final finish of the flooring tiles. Various stages of bamboo processing starting from cross-cutting, parallel splitting, knot removal, two-side planning, anti-fungal treatment, drying; four-side planning, glue application and hot pressing were fine tuned. Products such as flooring tiles, furniture sections, reconstituted wood, air locked sections, mat boards etc. have been developed under the project.

Composite Materials Towards Re-building and Rehabilitation

In the wake of disastrous damages by the earthquake in Gujarat, the Advanced Composites Programme has contributed to the national efforts of re-building and rehabilitation. Under the TIFAC Rehab Project for Kachchh, the following initiatives were taken up for the quake affected people:

- 392 low-cost semi-permanent shelters (20'×12') made of natural fibre composite materials such as jute-coir composite boards and rice husk particle boards with bamboo mat face veneer etc. were supported on MS angles & channels. For improved aesthetics and also to augment the thermal insulation, natural fibre composite board roofing of the shelters was covered with terracotta tiles.
- In order to cater to the shelters, 128 community toilet blocks (4'×4') made of modular FRP section for walls and roof were constructed.
- 15 shops (12'×8') were constructed in the township along with a Post Office in the township, which has commenced its services.
- In addition to the semi-permanent residential shelters constructed at Bhuj, 25 school blocks-cum-community centers (24'× 20') were also constructed at various locations in Kachchh.

The TIFAC Rehab Project was a model initiative of technology demonstration with novel building materials with the delivery in the quickest possible time addressing the crucial need for post-disaster relief.

Composite Building Materials – Technology Demonstration

For augmenting the reception block of Technology Bhavan campus in Delhi, the Advanced Composites Programme took initiatives by building a 3000 sq. ft. temporary structure for post office, CR section, and technology demonstration-cum-display area and additional office space towards showcasing composite building materials. The array of products developed under the programme such as jute-coir boards, FRP doors, bamboo composite flooring tiles and rice husk particle boards for false ceiling were used in the construction of the shelter towards technology demonstration.

Jute-coir composite boards made of coir felt and waste rubber-wood as inside veneers and oriented jute as face veneer is a unique value-added application for agro-wastes and positioned as an effective wood substitute building material. While the shelter structure was fabricated out of standard steel sections, jute-coir boards were used for double-wall construction ensuring excellent thermal insulation. They were also used for roofing overlaid with terracotta tiles. Elegant looking bamboo composite tiles were used for the flooring. The door shutters made of sandwich panels of glass fibre reinforced polyester resin, have good aesthetic appeal with adequate mechanical strength and water resistance.

References

- Composite-Materials-for-Biomedical-Applications-A-Review-46037763: researchgate.net, Retrieved 23, February 2020
- Composite-materials-for-civil-engineering-applications: tuf-bar.com, Retrieved 08, July 2020
- Engineering-Applications-of-Composite-Materials-3688361: academia.edu, Retrieved 14, May 2020
- Electrical-and-electronics-applications-of-polymers-and-composites: polymerexpert.biz, Retrieved 26, March 2020
- Application-composites-development-electric-vehicles-sjef-van-breugel: linkedin.com, Retrieved 07, June 2020

Permissions

All chapters in this book are published with permission under the Creative Commons Attribution Share Alike License or equivalent. Every chapter published in this book has been scrutinized by our experts. Their significance has been extensively debated. The topics covered herein carry significant information for a comprehensive understanding. They may even be implemented as practical applications or may be referred to as a beginning point for further studies.

We would like to thank the editorial team for lending their expertise to make the book truly unique. They have played a crucial role in the development of this book. Without their invaluable contributions this book wouldn't have been possible. They have made vital efforts to compile up to date information on the varied aspects of this subject to make this book a valuable addition to the collection of many professionals and students.

This book was conceptualized with the vision of imparting up-to-date and integrated information in this field. To ensure the same, a matchless editorial board was set up. Every individual on the board went through rigorous rounds of assessment to prove their worth. After which they invested a large part of their time researching and compiling the most relevant data for our readers.

The editorial board has been involved in producing this book since its inception. They have spent rigorous hours researching and exploring the diverse topics which have resulted in the successful publishing of this book. They have passed on their knowledge of decades through this book. To expedite this challenging task, the publisher supported the team at every step. A small team of assistant editors was also appointed to further simplify the editing procedure and attain best results for the readers.

Apart from the editorial board, the designing team has also invested a significant amount of their time in understanding the subject and creating the most relevant covers. They scrutinized every image to scout for the most suitable representation of the subject and create an appropriate cover for the book.

The publishing team has been an ardent support to the editorial, designing and production team. Their endless efforts to recruit the best for this project, has resulted in the accomplishment of this book. They are a veteran in the field of academics and their pool of knowledge is as vast as their experience in printing. Their expertise and guidance has proved useful at every step. Their uncompromising quality standards have made this book an exceptional effort. Their encouragement from time to time has been an inspiration for everyone.

The publisher and the editorial board hope that this book will prove to be a valuable piece of knowledge for students, practitioners and scholars across the globe.

Index

www.ingramcontent.com/pod-product-compliance
Lightning Source LLC
LaVergne TN
LVHW080736210926
848485LV00003B/566